Thermoelectricity - Recent Advances, New Perspectives and Applications

Edited by Guangzhao Qin

Published in London, United Kingdom

IntechOpen

Supporting open minds since 2005

Thermoelectricity - Recent Advances, New Perspectives and Applications
http://dx.doi.org/10.5772/intechopen.87866
Edited by Guangzhao Qin

Contributors
Zhenzhen Qin, Wiqar Hussain Shah, Waqas Muhammad Khan, Jianwei Zhang, Gengyuan Liang, Ying Xiao, Bowei Lei, Wenshu Zhao, Kuldeep Chand Verma, Manpreet Singh, Jamil Ur Rahman, Gul Rahman, Soonil Lee, Yuri Bokhan, Sepideh Akhbarifar, Guangzhao Qin

First published in London, United Kingdom, 2022 by IntechOpen
IntechOpen is the global imprint of INTECHOPEN LIMITED, registered in England and Wales, registration number: 11086078, 5 Princes Gate Court, London, SW7 2QJ, United Kingdom
Printed in Croatia

British Library Cataloguing-in-Publication Data
A catalogue record for this book is available from the British Library

Additional hard and PDF copies can be obtained from orders@intechopen.com

Thermoelectricity - Recent Advances, New Perspectives and Applications
Edited by Guangzhao Qin
p. cm.
Print ISBN 978-1-83962-382-0
Online ISBN 978-1-83962-610-4
eBook (PDF) ISBN 978-1-83962-611-1

For EU product safety concerns:
IN TECH d.o.o., Prolaz Marije Krucifikse Kozulić 3, 51000 Rijeka, Croatia, info@intechopen.com or visit our website at intechopen.com.

Meet the editor

Guangzhao Qin is a Professor of Mechanical Engineering at Hunan University, China. He received a BS from Zhengzhou University, China, in 2011, an MSc from the University of the Chinese Academy of Sciences in 2015, and a Ph.D. from RWTH Aachen University, Germany, in 2018. He was recognized as a Vebleo Fellow, received the 2017 Chinese Government Award for Outstanding Self-financed Students Abroad, and graduated Summa Cum Laude from RWTH. His research interests are high-performance thermal management and energy transport and conversion at multi-time and multi-length scales. Dr. Quin has published more than sixty papers in scientific journals.

Contents

Preface

As we face the challenges of an energy crisis, it is crucial to seek out next-generation energy sources that are economical, sustainable (renewable), and clean (environmentally friendly). There are many types of energy that can be utilized in industry, such as nuclear energy, radiation solar energy, hydrogen energy, and so on. In most cases, all energy is converted into electricity before it can be utilized. Heat plays a key role in energy conversion and utilization. Heat is traditionally used to create electricity, but the process is complex and noisy and creates large amounts of waste. With the advantage of first-hand solid-state conversion to electrical power from thermal energy, especially from the reuse of waste heat, thermoelectricity has valuable applications in reusing resources, making it cost-effective. Compared to the traditional methods of converting heat to electricity, thermoelectric devices have many advantages, such as all-solid-state features, no mechanical moving parts, long lifetime, and more. Consequently, thermoelectricity has many potential applications. For example, based on nuclear radiation with a long half-life, thermoelectricity can supply electricity for space stations and spaceships in deep space. In addition, thermoelectricity can also supply electricity for flexible wearable devices. As such, thermoelectricity has received extensive attention and is expected to be helpful for combating environmental pollution and saving energy.

The wide applications of thermoelectrics in commercial and industrial fields have long been limited by thermoelectric conversion efficiency. The continuous improvement of thermoelectric performance and the strive to increase the power output under the same heat source are the key foci in thermoelectric technology. Many efforts have been dedicated to improving thermoelectric efficiency, which is a very challenging task because of the strongly correlated properties involved in thermoelectrics. Efforts to boost thermoelectric efficiency focus on two aspects. The first approach focuses on improving electric performance while not affecting thermal transport. The second approach focuses on suppressing thermal transport while not affecting electric properties. All methods for boosting thermoelectric efficiency include designing or searching for novel materials with excellent properties, especially low-dimensional materials, and regulating performance via techniques such as strain engineering, doping, nanostructuring, and so on. Significant progress has been made in the past few years and thermoelectric efficiency has been largely improved. Future studies are expected to improve thermoelectric efficiency even further.

This book presents a comprehensive overview of the progress made in thermoelectrics over the past few years with a focus on charge and heat carrier transport from both theoretical and experimental viewpoints. It also presents new strategies to improve thermoelectricity. In addition, the book discusses device physics and applications to guide the research community.

This book project has lasted more than two years and in that time many things have happened. I would like to acknowledge my wife and daughter for their support. Also, I would like to thank my sister for her helpful suggestions and contributions during the editing of this book.

Guangzhao Qin
State Key Laboratory of Advanced Design
and Manufacturing for Vehicle Body,
College of Mechanical and Vehicle Engineering,
Hunan University,
Changsha, China

Chapter 1

Introductory Chapter: Thermoelectricity – Recent Advances, New Perspectives, and Applications

Guangzhao Qin

1. Introduction

Seeking for next-generation energy sources is crucial when facing the challenges of an energy crisis. The energy is expected to be economical, sustainable (renewable), and clean (environment-friendly). There exist lots of energy that can be utilized in industry, such as nuclear energy, radiation solar energy, hydrogen energy, etc. Generally, electricity is the most direct type of energy available to human society. And in most cases, all the other energies are firstly converted into electricity and then can be utilized. During the conversion of energy, heat plays a key role as shown in **Figure 1**. For instance, the power stations get heat from coal or fuel burning or nuclear radiating, which are used to heat water to produce steam. Then, the electricity generators are driven. Thus, in addition to the electricity that can be directly utilized, heat is the most common type for energy conversion and utilization. However, the traditional utilization of heat is usually to drive the electricity generator by heating water, which will cause a large waste, and the process is complex with lots of mechanical moving parts and noise.

2. Thermoelectrics

With the advantage of first-hand solid-state conversion to electrical power from thermal energy, especially from the reuse of waste heat, thermoelectricity shows its valued applications in reusing resources for the low cost of operation [1]. Compared to the traditional manners of converting heat to electricity, thermoelectric devices have lots of advantages, such as all-solid-state feature, no mechanical moving part, long lifetime, etc. Consequently, thermoelectricity has lots of potential applications. For instance, based on the nuclear radiation with a long half-life, thermoelectricity can supply electricity for space stations and spaceships in deep space far from stellar [2]. Besides, thermoelectricity can also supply electricity for flexible wearable devices, devices in the body, IoT devices, etc [3, 4]. Thus, with the numerous applications in lots of fields, thermoelectricity has received extensive attention and is expected to be helpful for the crisis of the environment and the saving of energy.

However, the wide applications in commercial and industrial fields have long been limited by thermoelectric conversion efficiency. The thermoelectric efficiency is characterized by the dimensionless figure of merit $ZT = S^2\sigma T/\kappa$ [5, 6], where S is the Seebeck coefficient, σ is the electrical conductivity, T is the temperature, and κ

IntechOpen

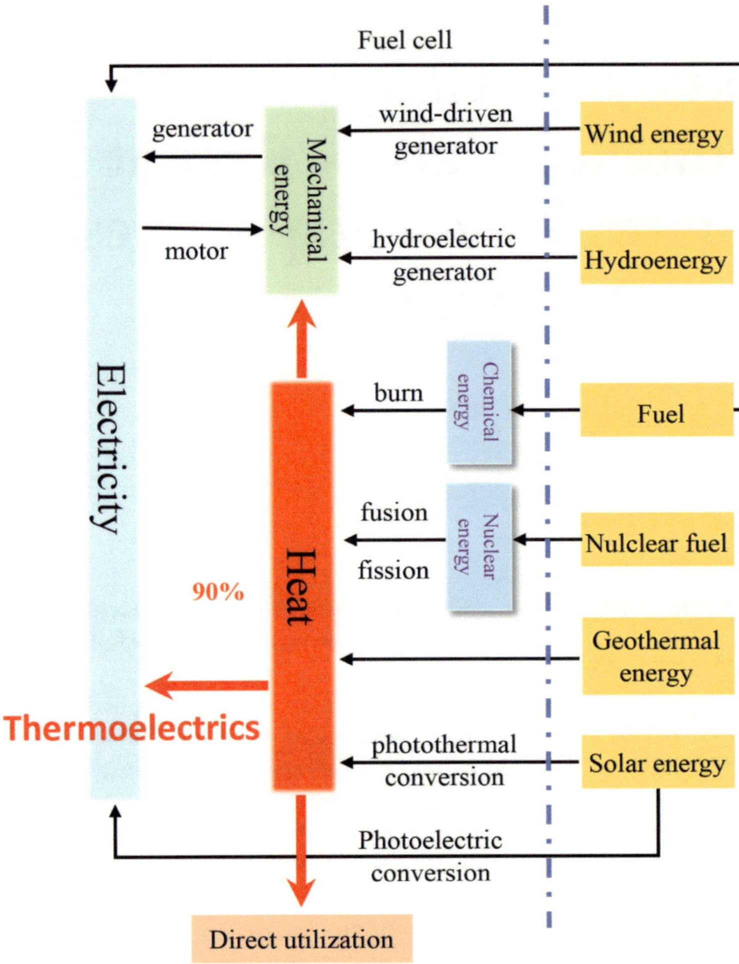

Figure 1.
The key role of heat in the energy conversion and the key role of thermoelectrics in the utilization of heat.

is the thermal conductivity contributed from both electron and phonon transport. The continuous improvement of thermoelectric performance and the strive to increase the power output under the same heat source are the key focus in thermoelectric technology. A lot of efforts have been dedicated to improving thermoelectric efficiency. However, it is a very challenging task because the parameters in the ZT formula are strongly correlated with each other. For instance, a large electrical conductivity is a benefit to the improvement of ZT, but a large thermal conductivity contributed from electron transport is simultaneously achieved, which plays a negative role in the improvement of ZT. Thus, the wide applications of thermoelectric devices are mainly limited by the low conversion efficiency from heat to electricity.

3. Improve thermoelectric performance

The efforts for boosting thermoelectric efficiency are mainly focused on two aspects. The first approach is to improve the electric performance while not affecting thermal transport. Another approach is to suppress the thermal transport while not affecting the electric properties. Moreover, the concept of "electron crystal & phonon glass" [7] has been proposed to improve the electric performance and to suppress the thermal transport simultaneously. All the manners for boosting thermoelectric efficiency include (1) designing or searching novel materials with excellent properties, especially for low-dimensional materials [1], and (2) regulating the performance using various manners such as strain engineering, doping, nanostructuring, etc. In addition, a new mechanism is also proposed for understanding the fundamentals of thermoelectrics. Significant progress has been made in the past few years, and thermoelectric efficiency has been largely improved. For instance, spin-related characteristics have been used to regulate the electron and phonon transport in magnetism materials and to achieve remarkable thermoelectric performance [8, 9]. Future studies are expected to further improve the thermoelectric efficiency following a similar procedure or finding a different way.

This book aims to comprehensively collect the progress made in thermoelectrics over the past few years with a focus on charge and heat carrier transport from both theoretical and experimental views. New perspectives are also presented for the challenges on the topics and for future studies, especially for the strategies to improve the thermoelectric figure of merit. Beyond that, new discussions on device physics and applications are also included to offer a guide to the research community.

Author details

Guangzhao Qin
State Key Laboratory of Advanced Design and Manufacturing for Vehicle Body, College of Mechanical and Vehicle Engineering, Hunan University, Changsha, P.R. China

*Address all correspondence to: gzqin@hnu.edu.cn

References

[1] Dresselhaus MS et al. New directions for low-dimensional thermoelectric materials. Advanced Materials. 2007; **19**:1043-1053

[2] Dubi Y, Di Ventra M. *Colloquium*: Heat flow and thermoelectricity in atomic and molecular junctions. Reviews of Modern Physics. 2011; **83**:131-155

[3] Akinwande D, Petrone N, Hone J. Two-dimensional flexible nanoelectronics. Nature Communications. 2014;**5**:5678

[4] Wan C et al. Flexible n-type thermoelectric materials by organic intercalation of layered transition metal dichalcogenide TiS_2. Nature Materials. 2015;**14**:622-627

[5] Qin G et al. Hinge-like structure induced unusual properties of black phosphorus and new strategies to improve the thermoelectric performance. Scientific Reports. 2014;**4**:6946

[6] Qin G, Hao K-R, Yan Q-B, Hu M, Su G. Exploring T-carbon for energy applications. Nanoscale. 2019; **11**:5798-5806

[7] Takabatake T, Suekuni K, Nakayama T, Kaneshita E. Phonon-glass electron-crystal thermoelectric clathrates: Experiments and theory. Reviews of Modern Physics. 2014;**86**:669-716

[8] Tian Q, Zhang W, Qin Z, Qin G. Novel optimization perspectives for thermoelectric properties based on Rashba spin splitting: A mini review. Nanoscale. 2021;**13**:18032-18043

[9] Peng C et al. Improvement of Thermoelectricity Through Magnetic Interactions in Layered $Cr_2Ge_2Te_6$. Physica Status Solidi (RRL) – Rapid Research Letters. 2018;**12**:1800172

Chapter 2

Optimization of Thermoelectric Properties Based on Rashba Spin Splitting

Zhenzhen Qin

Abstract

In recent years, the application of thermoelectricity has become more and more widespread. Thermoelectric materials provide a simple and environmentally friendly solution for the direct conversion of heat to electricity. The development of higher performance thermoelectric materials and their performance optimization have become more important. Generally, to improve the ZT value, electrical conductivity, Seebeck coefficient and thermal conductivity must be globally optimized as a whole object. However, due to the strong coupling among ZT parameters in many cases, it is very challenging to break the bottleneck of ZT optimization currently. Beyond the traditional optimization methods (such as inducing defects, varying temperature), the Rashba effect is expected to effectively increase the $S^2\sigma$ and decrease the κ, thus enhancing thermoelectric performance, which provides a new strategy to develop new-generation thermoelectric materials. Although the Rashba effect has great potential in enhancing thermoelectric performance, the underlying mechanism of Rashba-type thermoelectric materials needs further research. In addition, how to introduce Rashba spin splitting into current thermoelectric materials is also of great significance to the optimization of thermoelectricity.

Keywords: Thermoelectric materials, Rashba spin splitting, spin-orbit coupling, Seebeck coefficient

1. Introduction

Thermoelectric materials can use the Seebeck and Peltier effects to directly convert heat and electricity into each other [1–3], providing a simple and environmentally friendly solution for the direct conversion of heat to electricity, and is expected to play an important role in meeting future energy challenges effect. Thermoelectric equipment can not only directly convert the heat from the sun, radioisotopes, automobiles, industrial sectors and even the human body into electrical energy, but can also implement solid-state heat pumps based on electric drive for distributed refrigeration. In recent years, the application of thermoelectricity has become more and more widespread. The equipment based on thermoelectric materials has the advantages of miniaturization, quiet operation and no emission of greenhouse gases. The development of higher performance thermoelectric materials and their performance optimization have become more important.

The dimensionless thermoelectric figure of merit (*ZT* value) can be used as a key indicator to quantify thermoelectric performance, which is defined as, where *T* is the absolute temperature, *S* is the Seebeck coefficient, σ is the electrical conductivity, and are the electronic and lattice thermal conductivity, respectively. Improving the *ZT* value has always been the main goal of thermoelectric research. Generally, to improve the *ZT* value, electrical conductivity, Seebeck coefficient and thermal conductivity must be globally optimized as a whole object. In order to obtain a higher *ZT*, a high Seebeck coefficient *S*, a high electrical conductivity σ and a low thermal conductivity κ are required. However, these parameters have very strong coupling in many cases, and it is very challenging to increase the *ZT* value of typical thermoelectric materials. At the micro level, the high thermoelectricity of the material comes from the subtle coordination between bond covalent and ionicity, band polymerization and splitting, localization and divergence of electronic states, and the trade-off between carrier mobility and effective mass. Therefore, how to find a balance point, or even dig out a new *ZT* value optimization mechanism from a different perspective, is an important innovation basis for thermoelectric materials.

2. Common *ZT* optimization strategy

As the basic science of thermoelectric matures, the research of thermoelectric materials has also begun to develop rapidly. Among the materials reported with high thermoelectric properties, the *ZT* value of Bi_2Te_3 compound and its alloy form is about 1 at room temperature, which has been regarded as the highest standard of advanced thermoelectric materials in the thermoelectric field [4]. Until 1990, Hicks and Dresselhaus proposed that better thermoelectric performance could be designed through the "size effect", that is, to reduce the size [5, 6]. They found that Bi_2Te_3 with a quantum well (two-dimensional) or nanowire (one-dimensional)

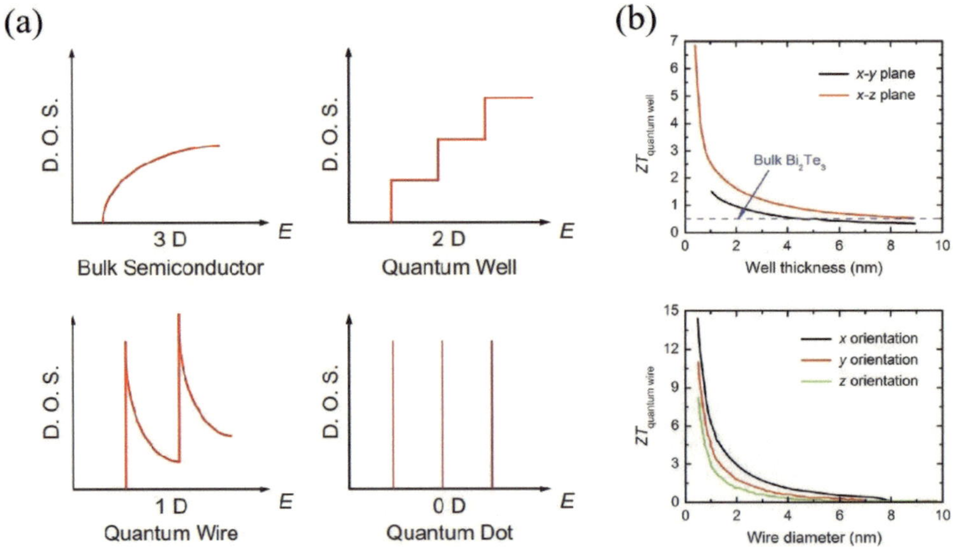

Figure 1.
(a) Schematic diagram of electronic density of states of three-dimensional bulk, two-dimensional quantum scale, one-dimensional nanoribbons, and zero-dimensional quantum dots; (b) the relationship between quantum-scale ZT *and layer thickness; the relationship between* ZT *and diameter of nanowires. The figure is taken from the literature [5–7].*

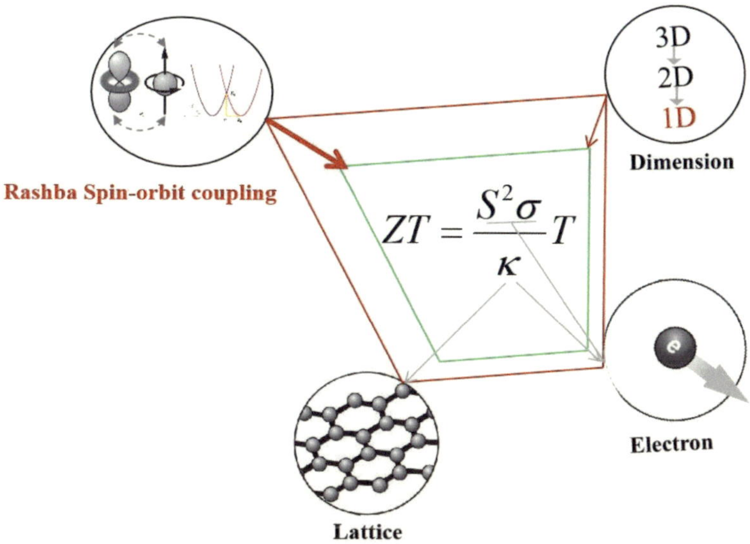

Figure 2.
Thermoelectric parameters can be affected by a combination of dimensions, charge, lattice, and spin-orbit coupling. Among them, the synergy among Seebeck coefficient S*, electrical conductivity* σ*, and thermal conductivity* κ *usually makes them as a whole to increase the* ZT *value, thereby realizing high-efficiency thermoelectric materials. In addition to reducing the size and dimensions, Rashba spin-orbit coupling may be a key direction for the development of next-generation thermoelectric materials.*

structure may have the potential to further increase the ZT value, and inferred that the root cause of the improvement in its thermoelectric performance is mainly due to the increased electronic density of states due to the decrease in dimensionality (**Figure 1a**), resulting in an increase in Seebeck coefficient. As shown in **Figure 1b**, the ZT value monotonously increases with the decrease of the thickness of the quantum well or the diameter of the quantum wire. Therefore, the use of size effect provides a new strategy for increasing the ZT value, because it not only increases the density of electronic states near the Fermi level E_F to increase the Seebeck coefficient, but also increases the phonon at the barrier-well interface. Boundary scattering reduces thermal conductivity.

The synergistic interaction between charge, lattice, and spin-orbit coupling is a necessary factor to further optimize thermoelectric performance (**Figure 2**). In the past three decades, people have discovered a variety of possible mechanisms that may affect transport performance: resonance energy levels, modulation doping, band convergence, classical and quantum size effects, anharmonicity, and spin-related effect, such as Rashba effect, spin Seebeck effect and topological state, etc., have been verified in various materials such as V_2-VI_3 compounds, V-VI group compounds, semi-Heusler alloys, diamond-like structure compounds and silicides. Among them, Rashba spin-orbit coupling and controllable anharmonicity may be the key to the development of next-generation thermoelectric materials.

3. ZT optimization based on Rashba spin splitting

3.1 The improved $S^2\sigma$ and reduced κ

Note that in the quantum size effect caused by the reduction of dimensionality, a key idea for optimizing the power factor $S^2\sigma$ is to increase the density of states (Density of States, DOS) near the Fermi level E_F. In the low-dimensional

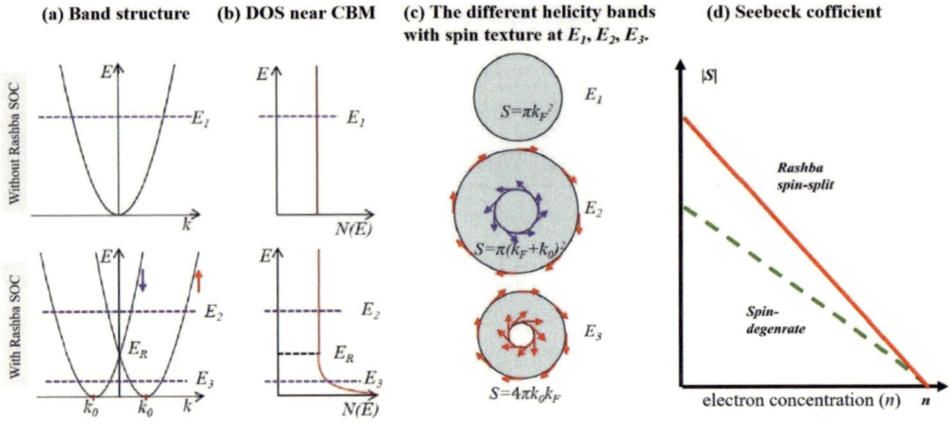

Figure 3.
(a) Rashba SOC leads to the spin splitting of the band structure; (b) the density of states diagram corresponding to figure (a) (near the bottom of the conduction band); (c) the energy levels E_1, E_2, E_3 reflect different spiral bands with spin texture on the Fermi surface. The red arrow represents spin up and the blue arrow represents spin down; (d) the Seebeck coefficient of Rashba spin split system and spin degenerate system varies with electron concentration n. the figure is taken from the literature [8].

Rashba material, DOS produces primitive sharp features near E_F due to spin splitting, which is very beneficial to the electron transport performance [7]. Due to the Rashba effect, the energy bands with different spin directions are split (as shown in **Figure 3a**), the density of states diagram will also show interesting characteristics, that is: when its DOS is located in the low energy region ($E_F < E_R$), it has a sharp peak (**Figure 3b**). The peak value of the low-energy DOS region is closely related to the band splitting amplitude k_0, that is, the larger the k_0, the sharper the DOS peak value. The increase in the density of states will lead to the dominance of ideal spin carriers, that is, a longer carrier lifetime can be maintained at room temperature. Rashba spin splitting can also cause spiral ribbons with different spin textures on the Fermi surface at different energy levels (**Figure 3c**). After considering the spin-orbit coupling (SOC) of Rashba, its power factor $S^2\sigma$ is almost twice that of spin degeneration. This huge enhancement effect is directly related to the special DOS contribution and the longer carrier lifetime caused by Rashba spin splitting. Therefore, when the Rashba SOC is included, the carriers near the band gap have a significant impact on the transmission performance, which plays a key role on obtaining higher $S^2\sigma$ and further optimization of the ZT value.

In addition, the giant Rashba spin splitting in a two-dimensional BiSb monolayer can increase the ZT value by a factor of two at room temperature compared to spin degenerate states [8]. Furthermore, it is confirmed that the Seebeck coefficient in Rashba spin-splitting BiTeX is higher than that in traditional spin-degenerate materials [9, 10], as shown in **Figure 3d**. Xiao et al. also predicted from the theoretical level that the Rashba system will exhibit abnormally enhanced thermoelectric behavior [11]. This is due to the Rashba SOC-induced scission of the density of states and extended carrier lifetime, and due to the relatively high DOS near the Fermi level. A large slope will result in a higher Seebeck coefficient, thereby increasing $S^2\sigma$. The idea that Rashba spin-orbit coupling can increase carrier lifetime has greatly stimulated more attempts in the Rashba system in photovoltaics and thermoelectrics [12]. At the level of phonon transport, Rashba SOC also affects the acoustic and optical branches of lattice vibration. For example, the long-wave optical LO mode will be strongly suppressed by the Rashba splitting energy band, resulting in a longer relaxation time [8, 12]. The intrinsic electric field in the Rashba system can also introduce non-intermittent phonons, so its lower thermal conductivity is also one of the main reasons for the high thermoelectric performance.

3.2 The link between Rashba and thermoelectric

Actually, the Rashba effect has attracted considerable attention in the fields of spintronics, ferroelectrics, and superconducting electronics [13, 14], Rashba spin-split generally originated from the SOC and inversion asymmetry, the SOC gives rise to a perturbing operator equal to $\lambda L \cdot S$ for electrons, where L and S are the total orbital and spin angular momenta, and λ the coupling constant [15]. The spin-orbital Hamiltonian has a Bychkov-Rashba form $H_{SOC} = \alpha_R(\sigma \times k) \cdot z$, where α_R is the Rashba parameter and represents the strength of the Rashba effect ($\alpha_R \propto \lambda E_z$), σ is the Pauli spin matrices, k is momentum, and z is the electric field direction along high-symmetry axis [16]. The bulk BiTeI with a giant Rashba effect has been confirmed by the angle-resolved photoemission spectroscopy and first-principles calculations [17, 18]. Interestingly, it is found that the heat-electricity converting thermoelectric effects are strongly influenced by the band structures near the Fermi level in BiTeI; thus the Rashba effect may also offer unusual opportunities for thermoelectrics [9].

As we known, thermoelectric materials are commonly composed of heavy elements with strong SOC [19]. In view of this, the spin-enabled mechanisms including the Rashba effect [20] and the spin Seebeck effect [21] offer new channels to manipulate and further optimize thermoelectric properties [22]. However, the spin Seebeck effect is currently limited at cryogenic temperatures. By contrast, the Rashba effect is promising to facilitate performance enhancement in broad thermoelectric materials. It was reported that Rashba spin splitting yielded a unique constant DOS near the E_F, which resulted in high S. On the other hand, the internal electric field induced by Rashba effect reinforced anharmonicity and introduced soft bonds reduced the κ_l [23]. Actually, the Rashba effect is likely to enlarge the band degeneracy for fulfilling a high-quality factor. In the non-centro symmetric materials, the strong SOC induces the Rashba effect where the original single band edge splits into two band extrema with energy shift and momentum offset [24]. Such spin splitting band provides a new way to engineer the band structure and enlarge the band degeneracy for enhancing thermoelectric performance. Thermoelectric materials with a phase transition may exhibit the Rashba effect due to the broken inversion symmetry. For instance, GeTe undergoes a phase transition from rhombohedral to cubic structures near 700 K and shows the giant Rashba effect [25, 26]. Therefore, the cubic-to-rhombohedral phase transition of GeTe provides a unique method to produce the Rashba effect in thermoelectric materials.

Recent classical strategies of quantum confinement effect [27], resonant level [28], band convergence [29], liquid-like ions [30], entropy engineering [31], anharmonicity [32], and modulation doping [33] have enhanced ZT in many thermoelectric material systems. Albeit the advances in thermoelectric theories, there's no doubt that the ZT enhancement have already reached a bottleneck. The Rashba effect, spin-dependent band splitting, has been proved to be a new path to enhance thermoelectric performance [34]. In detail, Hong et al. demonstrated a strong Rashba spin splitting in Sn-doped GeTe and results in the band convergence experimentally, so that $S^2\sigma$ is significantly enhanced. Additionally, the co-existence of stacking faults with other multiscale nanostructures yields an ultra-low thermal conductivity, thus achieved a high ZT over 2.2 [34]. A link between the Rashba effect and the enhancement of thermoelectric is well established. The demonstrated new strategy of exploring the Rashba effect to increase the band degeneracy for enhancing thermoelectric performance can provide guidelines to develop new-generation thermoelectric materials.

4. Conclusion

Although the Rashba effect has great potential in enhancing thermoelectric performance, the influence of Rashba spin-orbit coupling on various thermoelectric parameters, thermoelectric optimization rule and the exact mechanism are still to be explored to a large extent. In particular, the thermoelectric performance and the underlying mechanism of Rashba-type thermoelectric materials need further research, or how to introduce Rashba spin splitting into current excellent thermoelectric materials is also of great significance to the optimization of thermoelectricity.

Author details

Zhenzhen Qin
School of Physics and Microelectronics, Zhengzhou University, Zhengzhou, People's Republic of China

*Address all correspondence to: qzz@zzu.edu.cn

References

[1] DiSalvo, F. J. Thermoelectric Cooling and Power Generation. Science **1999**, *285* (5428), 703-706. https://doi.org/10.1126/science.285.5428.703.

[2] Bell, L. E. Cooling, Heating, Generating Power, and Recovering Waste Heat with Thermoelectric Systems. Science **2008**, *321* (5895), 1457-1461. https://doi.org/10.1126/science.1158899.

[3] Snyder, G. J.; Toberer, E. S. Complex Thermoelectric Materials. In *Materials for Sustainable Energy*; Co-Published with Macmillan Publishers Ltd, UK, 2010; pp 101-110. https://doi.org/10.1142/9789814317665_0016.

[4] D. M. Rowe. *CRC Handbook of Thermoelectrics*; CRC press.

[5] Hicks, L. D.; Dresselhaus, M. S. Thermoelectric Figure of Merit of a One-Dimensional Conductor. Phys. Rev. B **1993**, *47* (24), 16631-16634. https://doi.org/10.1103/PhysRevB.47.16631.

[6] Hicks, L. D.; Dresselhaus, M. S. Effect of Quantum-Well Structures on the Thermoelectric Figure of Merit. Phys. Rev. B **1993**, *47* (19), 12727-12731. https://doi.org/10.1103/PhysRevB.47.12727.

[7] Dresselhaus, M. S.; Chen, G.; Tang, M. Y.; Yang, R. G.; Lee, H.; Wang, D. Z.; Ren, Z. F.; Fleurial, J.-P.; Gogna, P. New Directions for Low-Dimensional Thermoelectric Materials. Adv. Mater. **2007**, *19* (8), 1043-1053. https://doi.org/10.1002/adma.200600527.

[8] Yuan, J.; Cai, Y.; Shen, L.; Xiao, Y.; Ren, J.-C.; Wang, A.; Feng, Y. P.; Yan, X. One-Dimensional Thermoelectrics Induced by Rashba Spin-Orbit Coupling in Two-Dimensional BiSb Monolayer. Nano Energy **2018**, *52*, 163-170. https://doi.org/10.1016/j.nanoen.2018.07.041.

[9] Wu, L.; Yang, J.; Wang, S.; Wei, P.; Yang, J.; Zhang, W.; Chen, L. Two-Dimensional Thermoelectrics with Rashba Spin-Split Bands in Bulk BiTeI. Phys. Rev. B **2014**, *90* (19), 195210. https://doi.org/10.1103/PhysRevB.90.195210.

[10] Wu, L.; Yang, J.; Chi, M.; Wang, S.; Wei, P.; Zhang, W.; Chen, L.; Yang, J. Enhanced Thermoelectric Performance in Cu-Intercalated BiTeI by Compensation Weakening Induced Mobility Improvement. Sci. Rep. **2015**, *5* (1), 14319. https://doi.org/10.1038/srep14319.

[11] Xiao, C.; Li, D.; Ma, Z. Unconventional Thermoelectric Behaviors and Enhancement of Figure of Merit in Rashba Spintronic Systems. Phys. Rev. B **2016**, *93* (7), 075150. https://doi.org/10.1103/PhysRevB.93.075150.

[12] Zheng, F.; Tan, L. Z.; Liu, S.; Rappe, A. M. Rashba Spin–Orbit Coupling Enhanced Carrier Lifetime in CH3NH3PbI3. Nano Lett. **2015**, *15* (12), 7794-7800. https://doi.org/10.1021/acs.nanolett.5b01854.

[13] Mathias, S.; Ruffing, A.; Deicke, F.; Wiesenmayer, M.; Sakar, I.; Bihlmayer, G.; Chulkov, E. V.; Koroteev, Yu. M.; Echenique, P. M.; Bauer, M.; Aeschlimann, M. Quantum-Well-Induced Giant Spin-Orbit Splitting. Phys. Rev. Lett. **2010**, *104* (6), 066802. https://doi.org/10.1103/PhysRevLett.104.066802.

[14] Cappelluti, E.; Grimaldi, C.; Marsiglio, F. Topological Change of the Fermi Surface in Low-Density Rashba Gases: Application to Superconductivity. Phys. Rev. Lett. **2007**, *98* (16), 167002. https://doi.org/10.1103/PhysRevLett.98.167002.

[15] Ziman, J. M. *Principle of the Theory of Solid*; Cambridge University Press: London, 1972.

[16] Bychkov, Yu. A.; Rashba, É. I. Properties of a 2D Electron Gas with Lifted Spectral Degeneracy. *Sov.* J. Exp. Theor. Phys. Lett. **1984**, *39*, 78.

[17] Ishizaka, K.; Bahramy, M. S.; Murakawa, H.; Sakano, M.; Shimojima, T.; Sonobe, T.; Koizumi, K.; Shin, S.; Miyahara, H.; Kimura, A.; Miyamoto, K.; Okuda, T.; Namatame, H.; Taniguchi, M.; Arita, R.; Nagaosa, N.; Kobayashi, K.; Murakami, Y.; Kumai, R.; Kaneko, Y.; Onose, Y.; Tokura, Y. Giant Rashba-Type Spin Splitting in Bulk BiTeI. Nat. Mater. **2011**, *10* (7), 521-526. https://doi.org/10.1038/nmat3051.

[18] Bahramy, M. S.; Arita, R.; Nagaosa, N. Origin of Giant Bulk Rashba Splitting: Application to BiTeI. Phys. Rev. B **2011**, *84* (4), 041202. https://doi.org/10.1103/PhysRevB.84.041202.

[19] Müchler, L.; Casper, F.; Yan, B.; Chadov, S.; Felser, C. Topological Insulators and Thermoelectric Materials. *Phys. Status Solidi RRL – Rapid* Res. Lett. **2013**, *7* (1-2), 91-100. https://doi.org/10.1002/pssr.201206411.

[20] Zhang, X.; Liu, Q.; Luo, J.-W.; Freeman, A. J.; Zunger, A. Hidden Spin Polarization in Inversion-Symmetric Bulk Crystals. Nat. Phys. **2014**, *10* (5), 387-393. https://doi.org/10.1038/nphys2933.

[21] Uchida, K.; Takahashi, S.; Harii, K.; Ieda, J.; Koshibae, W.; Ando, K.; Maekawa, S.; Saitoh, E. Observation of the Spin Seebeck Effect. Nature **2008**, *455* (7214), 778-781. https://doi.org/10.1038/nature07321.

[22] He, J.; Tritt, T. M. Advances in Thermoelectric Materials Research: Looking Back and Moving Forward. Science **2017**, *357* (6358). https://doi.org/10.1126/science.aak9997.

[23] Wu, L.; Yang, J.; Wang, S.; Wei, P.; Yang, J.; Zhang, W.; Chen, L. Two-Dimensional Thermoelectrics with Rashba Spin-Split Bands in Bulk BiTeI. Phys. Rev. B **2014**, *90* (19), 195210. https://doi.org/10.1103/PhysRevB.90.195210.

[24] Manchon, A.; Koo, H. C.; Nitta, J.; Frolov, S. M.; Duine, R. A. New Perspectives for Rashba Spin–Orbit Coupling. Nat. Mater. **2015**, *14* (9), 871-882. https://doi.org/10.1038/nmat4360.

[25] Zhang, X.; Bu, Z.; Lin, S.; Chen, Z.; Li, W.; Pei, Y. GeTe Thermoelectrics. Joule **2020**, *4* (5), 986-1003. https://doi.org/10.1016/j.joule.2020.03.004.

[26] Fahrnbauer, F.; Souchay, D.; Wagner, G.; Oeckler, O. High Thermoelectric Figure of Merit Values of Germanium Antimony Tellurides with Kinetically Stable Cobalt Germanide Precipitates. J. Am. Chem. Soc. **2015**, *137* (39), 12633-12638. https://doi.org/10.1021/jacs.5b07856.

[27] Zhu, T.; Liu, Y.; Fu, C.; Heremans, J. P.; Snyder, J. G.; Zhao, X. Compromise and Synergy in High-Efficiency Thermoelectric Materials. Adv. Mater. **2017**, *29* (14), 1605884. https://doi.org/10.1002/adma.201605884.

[28] Heremans, J. P.; Jovovic, V.; Toberer, E. S.; Saramat, A.; Kurosaki, K.; Charoenphakdee, A.; Yamanaka, S.; Snyder, G. J. Enhancement of Thermoelectric Efficiency in PbTe by Distortion of the Electronic Density of States. Science **2008**, *321* (5888), 554-557. https://doi.org/10.1126/science.1159725.

[29] Pei, Y.; Shi, X.; LaLonde, A.; Wang, H.; Chen, L.; Snyder, G. J. Convergence of Electronic Bands for High Performance Bulk Thermoelectrics. Nature **2011**, *473* (7345), 66-69. https://doi.org/10.1038/nature09996.

[30] Liu, H.; Shi, X.; Xu, F.; Zhang, L.; Zhang, W.; Chen, L.; Li, Q.; Uher, C.;

Day, T.; Snyder, G. J. Copper Ion Liquid-like Thermoelectrics. Nat. Mater. **2012**, *11* (5), 422-425. https://doi.org/10.1038/nmat3273.

[31] Liu, R.; Chen, H.; Zhao, K.; Qin, Y.; Jiang, B.; Zhang, T.; Sha, G.; Shi, X.; Uher, C.; Zhang, W.; Chen, L. Entropy as a Gene-Like Performance Indicator Promoting Thermoelectric Materials. Adv. Mater. **2017**, *29* (38), 1702712. https://doi.org/10.1002/adma.201702712.

[32] Zhao, L.-D.; Lo, S.-H.; Zhang, Y.; Sun, H.; Tan, G.; Uher, C.; Wolverton, C.; Dravid, V. P.; Kanatzidis, M. G. Ultralow Thermal Conductivity and High Thermoelectric Figure of Merit in SnSe Crystals. Nature **2014**, *508* (7496), 373-377. https://doi.org/10.1038/nature13184.

[33] Pei, Y.-L.; Wu, H.; Wu, D.; Zheng, F.; He, J. High Thermoelectric Performance Realized in a BiCuSeO System by Improving Carrier Mobility through 3D Modulation Doping. J. Am. Chem. Soc. **2014**, *136* (39), 13902-13908. https://doi.org/10.1021/ja507945h.

[34] Hong, M.; Lyv, W.; Li, M.; Xu, S.; Sun, Q.; Zou, J.; Chen, Z.-G. Rashba Effect Maximizes Thermoelectric Performance of GeTe Derivatives. Joule **2020**, *4* (9), 2030-2043. https://doi.org/10.1016/j.joule.2020.07.021.

Chapter 3

Processing Techniques with Heating Conditions for Multiferroic Systems of $BiFeO_3$, $BaTiO_3$, $PbTiO_3$, $CaTiO_3$ Thin Films

Kuldeep Chand Verma and Manpreet Singh

Abstract

In this chapter, we have report a list of synthesis methods (including both synthesis steps & heating conditions) used for thin film fabrication of perovskite ABO_3 ($BiFeO_3$, $BaTiO_3$, $PbTiO_3$ and $CaTiO_3$) based multiferroics (in both single-phase and composite materials). The processing of high quality multiferroic thin film have some features like epitaxial strain, physical phenomenon at atomic-level, interfacial coupling parameters to enhance device performance. Since these multiferroic thin films have ME properties such as electrical (dielectric, magneto-electric coefficient & MC) and magnetic (ferromagnetic, magnetic susceptibility etc.) are heat sensitive, i.e. ME response at low as well as higher temperature might to enhance the device performance respect with long range ordering. The magne-toelectric coupling between ferromagnetism and ferroelectricity in multiferroic becomes suitable in the application of spintronics, memory and logic devices, and microelectronic memory or piezoelectric devices. In comparison with bulk multiferroic, the fabrication of multiferroic thin film with different structural geometries on substrate has reducible clamping effect. A brief procedure for multiferroic thin film fabrication in terms of their thermal conditions (temperature for film processing and annealing for crystallization) are described. Each synthesis methods have its own characteristic phenomenon in terms of film thickness, defects formation, crack free film, density, chip size, easier steps and availability etc. been described. A brief study towards phase structure and ME coupling for each multiferroic system of $BiFeO_3$, $BaTiO_3$, $PbTiO_3$ and $CaTiO_3$ is shown.

Keywords: thin films synthesis, magnetoelectric coupling, film-on-substrate geometry

1. Introduction

Multiferroics have simultaneous ferroelectric and magnetic ordering and exhibits unusual physical properties in the sense of heat transport phenomenon at low and higher temperature turns to identify new device applications of spintronics due to their coupling between dual order parameters [1–6]. The first magnetoelectric (ME) effect studied by Dzyaloshinskii on Cr_2O_3 in 1960s was discussed in the report [2]. After few decades, the study has been taken on bulk composites of magnetostrictive ferrites and piezoelectric $BaTiO_3$. But the research was halted during number of years due to observed weak value of ME coupling. Now around 2003, the spin-dependent multiferroicity and strong ME effect in $TbMnO_3$ has been reported and the coexistence of antiferromagnetism and ferroelectric polarization in $BiFeO_3$ (BFO) have created much interest in multiferroics [3]. Such perovskite BFO sure multiferroicity due to the fact that magnetism requires unpaired d^n cations, while ferroelectricity due to d^0 configuration.

1.1 Mechanism of multiferroicity (heat sensitive electric and magnetic behavior) occur in perovskites structure

The ME effect in multiferroic is caused due to switching of magnetization M with an applied electric field E and vice-versa. Moreover, multiferroic switching states should remain (a "persistent" switch, not a transient), and be fast [6]. In single phase multiferroics, the ferroelectricity and ferromagnetism created due to same structural arrangement in ABO_3 like $BiFeO_3$, $BaTiO_3$, $PbTiO_3$, $CaTiO_3$ etc., *i.e.* either due to same ion (e.g. Fe^{3+}) and or the ferroelectricity arises due to one ion (e.g. the unpaired electron in Pb or Bi), while the magnetism via a second ion (e.g. Fe in $BiFeO_3$). For memory devices, it is desired that the multiferroic to be highly insulating and function at room temperature with a large switching charge. For a MERAM, it is likely to combine the ultrafast (250 ps) electrical WRITE operation with the non-destructive (no reset) magnetic READ operation, *i.e.* combining the best qualities of FRAM and MRAM. However, the magnetization may small for an effective READ.

1.2 Multiferroic heterostructure

The multiferroic heterostructures is actually the thin film form of the multiferroic. For this, **Figure 1** shown the electric field control and switch of the local magnetism which has two coupling mechanisms exists in ferromagnet-multiferroic heterostructure [4]. In addition to the macroscopic stress, the heteroepitaxy (the growth of a material B on a different material A) caused an intrinsic epitaxial stress so that the crystal symmetry, lattice constant, and/or chemical bonds do not matched perfectly. During the start of the films growth, the substrate lattice constants might to form stresses. Besides to the single-phase multiferroic such as $BiFeO_3$, $BaTiO_3$, $PbTiO_3$, $CaTiO_3$, the composite multiferroic heterostructures constructed in the way to realize an artificial systems in the form of thin film. Wang *et al.* [3] reported an enhancement of polarization in heteroepitaxially constrained $BiFeO_3$ thin film which display a room-temperature spontaneous polarization (50 to 60 $\mu C\ cm^{-2}$) that is much higher (in magnitude) than from bulk (6.1 $\mu C\ cm^{-2}$). The films thickness is also play an important role to turn multiferroicity in heterostructure.

1.3 Strain mediated magnetoelectric effects

The ME coupling in two-phase multiferroics is the strain transfer phenomenon occurs among two phases which is schematically constructed for 1–3 nanocomposite (**Figure 2**) [5]. It is the polarization (magnetization) that changed by a magnetic (electric) field. The ME effect in composites is a product tensor property of the magnetostrictive or magnetoelastic effect (magnetization and lattice strain coupling) in one phase and the piezoelectric effect (polarization and lattice strain

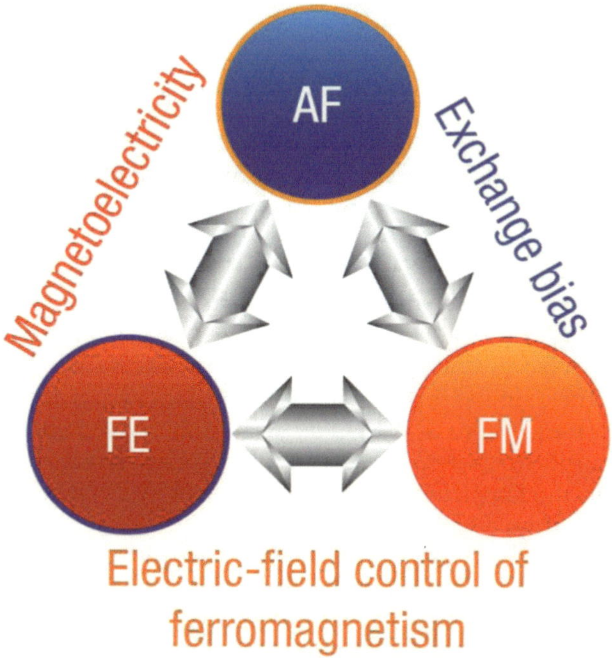

Figure 1.
Representation (schematic) of electric control of magnetism due to existence of ferroelectricity, antiferromagnetism and ferromagnetism [4].

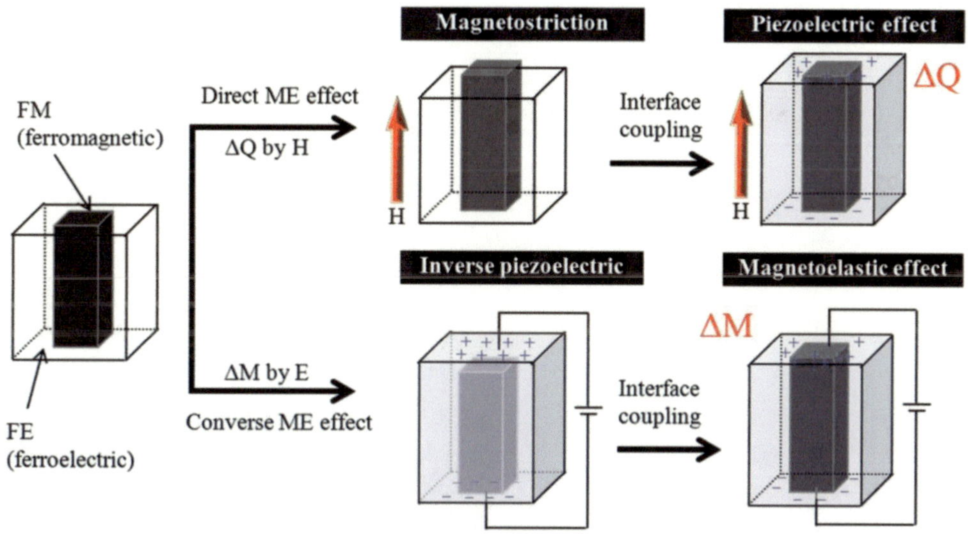

Figure 2.
Strain-mediated ME coupling in composites of a ferromagnetic (FM) and a ferroelectric (FE) phases (ΔQ: induces surface charges and ΔM: induces a magnetization change or domain reorientation) [5].

coupling) in the other phase. This ME coupling is the measurement due to direct or converse ME coefficient which detects an electrical signal with applied magnetic field. The strain generated in the magnetostrictive phase by a magnetic field induces surface charges in the piezoelectric phase. The direct ME voltage coefficient is given by $\alpha_E = \Delta E/\Delta H_{ac}$.

2. Multiferroic perovskite thin films and their processing techniques

2.1 $BiFeO_3$

Recently, the multiferroic systems such as $BaTi_2O_4$, $YMnO_3$, $BiMnO_3$, $LuFe_2O_4$, and $BiFeO_3$ have been widely investigated [7]. Among them, $BiFeO_3$ (BFO) with high $T_c \sim 1103$ K and $T_N \sim 643$ K attracts much attention due to its simultaneous ferroelectric and antiferromagnetic behaviors exist even at room temperature. This BFO has ferroelectricity occurs due to $6s^2$ lone pair of electrons of Bi^{3+} where structural distortion take-place and the magnetism occurs via superexchange interactions in Fe-O-Fe ions [8, 9]. This suggested to the polarization enhancement in BFO via chemical substitution along A site from rare-earth such as La^{3+}, Sm^{3+}, and Dy^{3+} ions due to their similar ionic radius and isovalent chemistry with Bi. This substitution of rare-earth into Bi in BFO may also induce a reduction in T_c value and formation of an antiferroelectric phase, but had minimal effect on antiferromagnetism. Moreover, the magnetization in BFO resulted by G-type antiferromagnet order is a cycloidal of wavelength, $\lambda \sim 62–64$ nm [10].

2.1.1 Multiferroic $BiFeO_3$ thin film

The bulk $BiFeO_3$ is a room-temperature ferroelectric with a spontaneous electric polarization directed along one of the [111] axes in the perovskite structure as shown in **Figure 3a** [11]. The ferroelectricity due to lattice distortions reduces the symmetry from cubic to rhombohedral to cause ferroelastic strain. With applied an electric field, the ferroelectric polarization in $BiFeO_3$ have eight possible positive

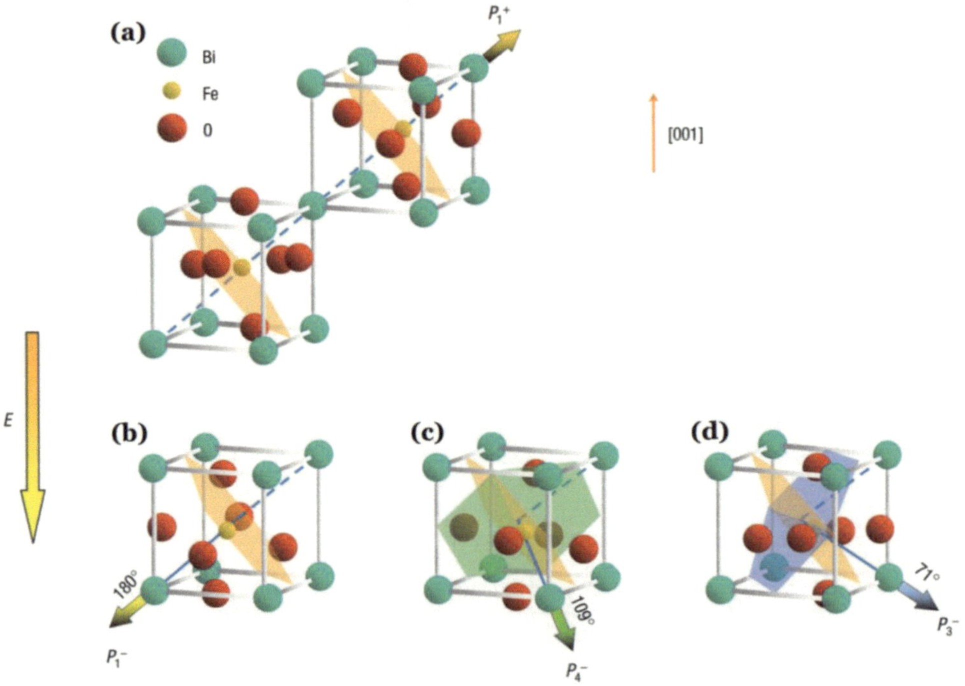

Figure 3.
The (001)-oriented $BiFeO_3$ crystal structure (schematic) and the ferroelectric polarization (bold arrows) and antiferromagnetic plane (shaded planes). (a) Polarization with an up out-of-plane component before electrical poling. (b) 180° polarization switching with the out-of-plane component switched by an electrical field (no change in antiferromagnetic phase). 109° (c) and 71° (d) polarization switching with the out-of-plane by an electrical field (antiferromagnetic plane changes) [11].

and negative orientations along the four cube diagonals, and the direction of the polarization can be switched by 180°, 109° and 71° as shown in **Figure 3b–d**. Switching of the polarization by either 109° or 71° changes the rhombohedral axis in the lattice system due to switching of the ferroelastic domain state. The antiferromagnetic ordering is G-type for which the nearest neighbor Fe moments aligned antiparallel to each other. In bulk $BiFeO_3$, the orientation of the antiferromagnetic vector has a long-wavelength spiral, which may suppress in thin films [12]. The DFT first principle calculations suggested the preferred orientation of the individual spins, in the absence of the spiral, which is perpendicular to the rhombohedral axis. It results into polarization switching by either 71° or 109° that should change the orientation of the easy magnetization plane. **Figure 1d** is the ferroelectric switching with 71° which leads to a reorientation of the antiferromagnetic order results into ME switching effects expectable in BFO films.

2.1.2 Synthesis techniques for $BiFeO_3$ thin films

In **Table 1**, we have included a lot of study to summarize chemical synthesis routes for processing of multiferroic $BiFeO_3$ thin films. The observed values of grain size, thickness and quality of film might be depends upon material processing and film coating technique.

2.2 $BaTiO_3$

$BaTiO_3$ (BTO) has shown multiferroicity at room temperature [36]. The three lattice structural phases of BTO are ferroelectric: rhombohedral < 190 K, orthorhombic for 190 K < T < 278 K and tetragonal for 278 K < T < 395 K. The paraelectric BTO phase exists at higher temperature. With tetragonal BTO, $a = b = 0.399$ nm and $c = 0.403$ nm of P4mm space group. The spontaneous electric polarization of BTO lattice might be related with the displacement of the Ti^{4+} ion along the c-axis which is sensitive to the charge in hybridization of the 3d states of the Ti cation with the 2p states of the surrounding O anions. For single-phase $BaTiO_3$ with doping of transition metal ions, the Fe-doping into BTO is widely reported due to excellent magnetic behavior that originated from unpaired spin of Fe^{3+} ions [37]. However, the multiferroic nanocomposites may selected from perovskites such as $BaTiO_3$, $PbTiO_3$ and $BiFeO_3$ (due to their large polarization and piezoelectric coefficients) with ferrites which have high magnetostriction value, high resistivity, T_N above room temperature and large ferromagnetism [38].

2.2.1 Multi-layered heteroepitaxial multiferroic thin films

The ME coupling in composite phases is effectively tuned through interfacial strain, exchange-bias, field effects, and so on [39]. It is a way to construct film geometry on the basis of their dimensions. For example, a 0–3 configuration means that there are two phases in the composite, one consisting of zero-dimension particulates, and the other is three-dimensional bulk. However, the ME composites for film-on-substrate, the multi-layered epitaxial thin films (2–2 configuration) were firstly reported that to be results into strong ME coupling with higher quality of crystallography and intimate coherent interface. But the clamping effect of the substrate onto the ferroelectric (FE) phase reduces the magnitude of the ME coupling coefficient in thin films. For 1–3 configuration, *i.e.* the vertical nanostructures from perovskite and spinel systems, like $CoFe_2O_4$ nanopillars embedded in a $BaTiO_3$ or $BiFeO_3$ matrix, the substantial enhancement of ME coupling and electric-field

Synthesis method	Chemical composition; reaction time; precursor salts	Brief synthesis procedure & heating conditions of processing	Substrate; grain size (x); film thickness (d)
Sol–gel & spin coating [13]	**$BiFeO_3$:Ti, Sm, Nd;** Hours; $BiC_6H_5O_7$, $FeC_6H_5O_7.H_2O$, Sm_2O_3, Nd_2O_3, $Ti(iOPr)_4$, NH_3, C_2H_7NO, $C_6H_8O_7$, H_2O_2	Fe(III) and Bi(III) aqueous solution prepared by required amount of metal citrates in water. Fe(III) solution kept at 80°C with pH below 1.5 and added ammonia and ethanolamine into Bi(III) to increase the solution stability. The Sm & Nd salts mixed in citric acid & water and refluxed overnight at 120°C to get solid citrate and add NH_3 while refluxing at 110°C/2 h. The Ti(IV) aqueous prepared by hydrolyzing Ti (IV)-isopropoxide to precipitate the Ti(IV) hydroxide and the addition of a mixture of citric acid and hydrogen peroxide with heating at 60°C dissolves the precipitated hydroxide. Final solution dissolved with controlling pH and spin coated 3000 rpm/30s onto Pt(111)Ti/SiO_2/Si substrate and annealed at 600°C/ 1 h.	Pt(111)Ti/SiO_2/Si; $x < 100$ nm; $d \sim 250$ nm
Atomic layer deposition (**ALD**) [14]	**$BiFeO_3$;** Hours; tris(2,3-dimethyl-2-butoxy)bismuth(III), iron (III) tert-butoxide	Thin films of Bi-Fe-O and individual BiO_x and FeO_x grown by ALD on Pt/ SiO_2/Si wafers using a commercial ALD reactor and precursors heated to 135–145°C providing enough vapor pressure for the deposition. The ozone (O_3) used as an oxidizing agent. The pulse duration for Bi $(mmp)_3$ and for ferrocene each <1 s and for ozone 5 s. The transport precursors kept at 150°C. The substrate placed ~3 cm from the gas inlet and the chamber heated to 250° C, and the gas outlet line kept at 100–150°C.	Pt/SiO_2/Si; x = 30 nm; d = 50 nm
Chemical spray [15]	**$BiFeO_3$:W;** Hours; $Bi(NO_3)_3{\cdot}5H_2O$, Fe $(NO_3)_3{\cdot}9H_2O$, $C_6H_8O_7$, Na_2WO_4	The Bi, Fe salts & 0.2 M citric acid dissolved in distilled water (added nitric acid to dissolve $Bi(NO_3)_3{\cdot}5H_2O$ and a 15 ml solution ready for spray. Second, 0.2 M $Na_2WO_4{\cdot}2H_2O$ added into a previous solution. These solutions sprayed separately onto non-conducting glass substrates at 400°C and the deposited films annealed at 500°C/ 4 h.	Non conducting glass; $x \sim 100$ nm; d = 1 μm
Chemical solution deposition using hydrothermal process [16]	**$BiFeO_3$;** Days; $Bi(NO_3)_3{\cdot}5H_2O$, Fe $(CH_3COCHCOCH_3)_3$, $C_2H_4O_2$, $C_3H_8O_2$, Bi $(NO_3)_3{\cdot}5H_2O$, Fe $(NO_3)_3{\cdot}9H_2O$, KNO_3, PVA, KOH	BFO nanopowders as seeds prepared by hydrothermal method. 0.005 M $Bi(NO_3)_3{\cdot}5H_2O$ and 0.005 M Fe $(NO_3)_3{\cdot}9H_2O$ dissolved in 100 ml of diluted HNO_3 (10%) and 12 M KOH solution slowly added to the above solution to adjust pH ~8. The precipitates filtered and washed and mixed with 30 ml KOH (12 M) and 15 ml PVA (4 g/l). The suspension	Flexible polyimide; $x \sim 70$ (seeded) and ~ 100 nm (seeded + UV BFO); $d \sim 200$ nm

Synthesis method	Chemical composition; reaction time; precursor salts	Brief synthesis procedure & heating conditions of processing	Substrate; grain size (x); film thickness (d)
		solution transferred into Teflon lined stainless-steel autoclaves and kept at 160°C/9 h. The obtained BFO powders dispersed (using ultrasonic processor) in acetic acid and 1,3-propanediol. The seeded photosensitive BFO precursors used direct fabrication of BFO films onto a polyimide substrate under UV irradiation to get crystallized.	
Aerosol Assisted chemical vapor deposition [17, 18]	**$BiFeO_3$**; Days; $[CpFe(CO)_2]_2$, $BiCl_3$, dichloromethane, THF	Aerosol assisted chemical vapor deposition (AACVD) uses a liquid–gas aerosol to transport soluble precursors to a heated substrate. A conventional atmospheric pressure CVD precursor proves in volatile or thermally unstable. For AACVD, the restrictions of volatility and thermal stability lifted the ionic precursors and metal oxide clusters used in aerosol assisted depositions. The AACVD reactions carried out using an in-house built coldwall CVD and annealed at 300°C	SICO coated float glass; $x \sim 100$ nm; d = 320–1700 nm
Metal–organic precursor complex solution [19]	**$BiFeO_3$**; **Hours**; $FeCl_3$, BCl_3, thiourea, methanol	The precursor solution prepared from ferric trichloride (0.1 M), bismuth trichloride (0.1 M), and thiourea (0.3 M, as an organic ligand) dissolved in methanol. The precursor complex is spin-coated onto the cleaned substrates at 1200 rpm and preheated at 473 K/10 min and annealed at 673 K/2 h.	Soda lime glass; $x \sim 150$ nm; d = 500 nm
Sol-electro-phoretic deposition [20]	**$BiFeO_3$**; **Hours**; $Bi(NO_3)_3{\cdot}5H_2O$, $Fe(NO_3)_3{\cdot}9H_2O$, 2-methoxy ethanol, citric acid, ethylene glycol	The solution prepared by dissolving $Bi(NO_3)_3{\cdot}5H_2O$ and $Fe(NO_3){\cdot}9H_2O$ in 2-methoxy ethanol. The citric acid and ethylene glycol added as complexing agents. The mixture sonicated for 10 min at room temperature until to gain a clear red sol suspension. The stainless steel mesh as working electrode and stainless steel (same dimension) as counter electrode and this substrate polished with emerged paper in oxalic acid and cleaned by acetone/ethanol. The EPD process for BFO film fabrication performed in red sol solution at 12 V within 60 s and the film crystalized at 500°C/2 h.	Stainless steel mesh; x = 100–150 nm; d = 196 nm
Sol–gel: Evaporation-induced self-assembly (using **Dip** Coating) [21]	**$BiFeO_3$**; Days; $Fe(NO_3)_3$, $Bi(NO_3)_3{\cdot}5H_2O$, 2-methoxyethanol, ethanol, glacial acetic acid	PB51-b-PEO62 block-copolymer with MW_{PB} = 51 000 g mol^{-1} and MW_{PEO} = 62 000 g mol^{-1} are polymer source as structure-directing agent. For solution A, PB51-b-PEO62 block-copolymer	$Pt/TiO_2/SiO_2/Si$; **Nanopores** (pore size = 100 nm); d = 66 nm

Synthesis method	Chemical composition; reaction time; precursor salts	Brief synthesis procedure & heating conditions of processing	Substrate; grain size (*x*); film thickness (*d*)
		dissolved in ethanol and 2-methoxyethanol at 70°C. Solution B prepared by dissolution of iron(III) nitrate and bismuth(III) nitrate in a mixture of 2-methoxyethanol, ethanol and glacial acetic acid. Solution B added to solution A to form the final solution. Nanopatterned porous and dense $BiFeO_3$ thin films deposited by dip-coating on $Pt/TiO_2/SiO_2/Si$ at 90°C and heated at 300°C/20 h, and annealed at 500°/10 min	
Hydro-thermal followed pulsed laser deposition [22]	**$BiFeO_3$**; Hours; $Bi(NO_3)_3.5H_2O$, $FeCl_3.6H_2O$, KOH	In hydrothermal process, 0.61 mM bismith nitrate and 0.55 mM iron chloride mixed with 100 mL of distilled water, and 0.9 M of KOH added as a mineralizer. The solution transferred to a 25-mL Teflon-lined autoclave and SRO/STO substrate positioned face down and placed ~1 cm above the bottom of the Teflon liner using Pt wire. A 10-nm-thick SRO layer deposited onto a STO substrate as the bottom electrode using pulsed laser deposition (PLD). The thin films deposited using a 4ω Nd:YAG laser (266 nm, 5 Hz repetition rate), the laser beam on ceramic targets kept to ~1.5 J/cm^2. The hydrothermal reaction taken at 200°C/6 h.	$SrRuO_3/SrTiO_3$; $x \sim 30$ nm $d = 60$ nm
Lithographic technique [23]	**0.65 BFO-0.35CFO**; Hours	Target material of BFO-CFO is prepared by any of the chemical method. BFO-CFO self-assembled nanostructures prepared by pulsed laser deposition using a 248 nm KrF laser. Growth carried out at 700°C in an oxygen atmosphere (150 mTorr) with a laser energy density of 3 J/cm^2 at 10 Hz. Etching of the annealed BFO-CFO performed in dilute HCl (50%, v/v) for 1 h at room temperature.	$SrTiO_3$; $x \sim 100$ nm; $d = 500$ nm
Solid state reaction - Pulsed laser deposition [24]	**$BiFeO_3$**; Days; Bi_2O_3, Fe_2O_3, isopropyl alcohol	High-purity Bi_2O_3 and Fe_2O_3 powders weighed with 10 mol% excess Bi and thoroughly mixed by ball milling for 15 h using high-purity isopropyl alcohol and the mixture dried and calcined at 500–800°C/1.5 h. The leached residue of calcined powder pressed into pellets and sintered at 730°C/1 h. BFO thin films grown by PLD. A $SrRuO_3$ electrode having 50 nm thickness deposited on a $SrTiO_3$ (100) substrate at 600°C	$SrTiO_3$; $x \sim 150$–200 nm $d \sim 250$ nm

Synthesis method	Chemical composition; reaction time; precursor salts	Brief synthesis procedure & heating conditions of processing	Substrate; grain size (x); film thickness (d)
		in an oxygen ambient of 100 mTorr.	
Microwave assisted sol–gel & spin coating [25]	**$BiFeO_3$**; Days; $Fe(NO_3)_3{\cdot}9H_2O$, $Bi(NO_3)_3{\cdot}5H_2O$, $(CH_2OH)_2$	Bismuth nitrate and Iron nitrate dissolved in ethylene glycol at room temperature separately and then mixed together. Resultant solution subjected to microwave radiations using microwaves operated at 2.45 GHz (microwave powers in 180–1000 W). The microwave assisted bismuth iron oxide solution spin coated on Cu substrates with using 3000 rpm/20s and then annealed in vacuum and 500 Oe field at 300°C/60 min.	Cu; x = 16–26 nm d = 700 nm
Non-aqueous sol–gel [26]	**$Bi_{0.9}Ho_{0.1}FeO_3/TiO_2$**; Days; $Bi(NO_3)_3.5H_2O$, $Fe(NO_3)_3.9H_2O$, $Ho(NO_3)_3.5H_2O$, $C_6H_8O_7$, $C_2H_6O_2$, $Ti(O\text{-}nBu)_4$, n-BuOH, $C_5H_8O_2$, CH_3COOH	For 1 g $Bi_{0.9}Ho_{0.1}FeO_3/TiO_2$ (BHFO) powder, the stoichiometric proportion of $Bi(NO_3)_3.5H_2O$ (0.003 M), $Fe(NO_3)_3.9H_2O$ (0.0033 M), $Ho(NO_3)_3.5H_2O$ (0.00033 M), $C_6H_8O_7$ (0.0067 M) and $C_2H_6O_2$ (10 ml) dissolved in deionized water and the solution heated at 75–85°C/4 h to get gel which dried at 100°C/24 h and annealed at 500°C/2 h. For the preparation of BHFO nanoparticles/TiO_2 composite thin films, first n-BuOH (0.0884 M) and $C_5H_8O_2$ (0.0015 M) mixed and, then $Ti(O\text{-}nBu)_4$ (0.005 M) added to the solution. CH_3COOH (0.001 M) slowly added into the alkoxide solution and stirred. Thin films containing 5 mol.% BHFO ($T^{95}B^5$), 10 mol.% BHFO ($T^{90}B^{10}$) and 20 mol.% BHFO ($T^{80}B^{20}$) prepared using non aqueous sol–gel method. To synthesize composite films, molar amount of n-BuOH, $C_5H_8O_2$, $Ti(O\text{-}nBu)_4$ and CH_3COOH kept similar as that of pure thin film. Firstly 0.0442 M of n-BuOH taken and calculated amount of BHFO nanoparticles mixed into it for every composite film. The whole mixture dispersed vigorously in an ultrasonic bath and then $C_5H_8O_2$, $Ti(O\text{-}nBu)_4$ and rest of n-BuOH (0.0442 M) added into this mixture and then taken into ultrasonic bath in which CH_3COOH added dropwise. The solutions spin-coated onto the glass substrate at 2000 rpm/30s and annealed at 500°C/2 h.	Glass; x ~ 60 nm; d = 200 nm (TiO_2), 244 nm ($T^{95}B^5$), 477 nm ($T^{90}B^{10}$), 635 nm ($T^{80}B^{10}$)
Photosensitive sol–gel using	**$Bi_{0.85}La_{0.15}Fe_{0.95}Mn_{0.05}O_3$**; Days; $Bi(NO_3)_3.5H_2O$, La	Benzoylacetone BzAcH preferred as the chelating agent and 2-methoxyethanol (MOE) as the	Si (100); x ~ 50 nm d = 140 nm

Synthesis method	Chemical composition; reaction time; precursor salts	Brief synthesis procedure & heating conditions of processing	Substrate; grain size (x); film thickness (d)
Dip coating [27]	$(NO_3)_3.6H_2O$, Mn $(NO_3)_2.4H_2O$, Fe $(NO_3)_3.5H_2O$, Benzoylacetone,	solvent. Four kinds of solutions of bismuth nitrate-BzAcH-MOE, lanthanum nitrate-BzAcH-MOE, manganese nitrate-BzAcH-MOE and ferric nitrate-BzAcH-MOE with the same molar ratio of nitrate: BzAcH:-MOE = 1:0.6:4 prepared beforehand. Therefore, the BLFMO solution obtained by mixing these four solutions with the total metallic ion concentration of 0.5 mol L^{-1}. The BLFMO gel film fabricated on Si (100) substrate by the dip-coating, and undergo irradiation and annealed at 650°C/1 h.	
Polymer assisted deposition from aqueous solutions [28]	**$BiFeO_3$**; Days; $Bi(NO_3)_3.5H_2O$, Fe $(NO_3)_3.9H_2O$, ethylenediaminetetraacetic acid, polyethylenimine	Heteroepitaxial thin films of BFO (a^{pc}_{BFO} = 3.965 Å) deposited on cubic (001)STO (a_{STO} = 3.905 Å) by Polymer assisted deposition from aqueous-based solutions. Individual solutions of Bi^{3+} and Fe^{3+} prepared by dissolving their respective hydrated nitrates in deionized water with ethylene diamine tetra acetic acid (EDTA, 1:1 molar ratio) and polyethylenimine (PEI) (2:1 and 1:1 mass ratio to EDTA for Bi and Fe, respectively). The corresponding pH was 8.1 for Bi and 5.1 for Fe. Each single solution then filtrated and the retained portions analyzed by inductively coupled plasma atomic emission spectroscopy yield a final concentration of 79 and 200 mM for Bi and Fe, respectively. The filtrated solutions mixed according with final stoichiometry and concentrated to a final value 0.25–0.3 M, in order to obtain films in the range 25–30-nm-thick. 30 μL of final solution with Bi/Fe (1:1) was spin coated (4000 rpm/20 s) on (001)STO substrates and annealed in 600–900°C/3 h.	(001)STO; $x \sim 30$ nm d = 30–60 nm
Radical enhanced atomic layer deposition (**RE-ALD**) [29]	**$BiFeO_3$**; Days; Tris(2,2,6,6-tetramethyl-3,5-heptanedionato) bismuth(III) [Bi $(TMHD)_3$], $Fe(TMHD)_3$	The deposition of metal oxide thin films used β-diketonate precursors and oxygen radicals. The multibeam system used in this study consists of a 10 in.-outer diameter stainless steel main chamber along with a load-lock assembly to facilitate sample insertion and removal without exposing the entire system to atmosphere. Pressure in the chamber maintained between 1×10^{-6} Torr at base and 2×10^{-5} Torr during operation by a CTI 4000 L/s cryogenic pump. The beams of the atom source and the precursor dosers	$(001)SrTiO_3$; x = 50–100 nm d = 93.5 nm

Synthesis method	Chemical composition; reaction time; precursor salts	Brief synthesis procedure & heating conditions of processing	Substrate; grain size (x); film thickness (d)
		converged onto a heated sample stage where the substrate mounted. Oxygen radicals produced from the atom source using a 2.46 GHz Sairem microwave power supply at 25 W and ∼0.6 sccm O_2 gas. Depositions carried out from 190 to 230°C.	
RF magnetron sputtering [30]	**$BiFeO_3$-$CoFe_2O_4$**; Days; Bi_2O_3, Fe_2O_3, Co_3O_4	BFO-CFO nanocomposite ceramic target prepared by a conventional oxide sintering method from Bi_2O_3 and Fe_2O_3. The final composite target prepared from BFO and CFO powders and mixed using ball milling for 24 hours then dried and pressed into a pellet and sintered at 800°C/5 h. The 2 inch diameter sintered target polished then bonded to a Cu plate using indium shot on a hot plate at 200°C. BFO-CFO nanocomposite films grown on (001)Nb-doped $SrTiO_3$ substrates. Deposition conducted at substrate temperatures of 480-650°C and working pressure 50 mTorr with Ar: oxygen ratios of 1:4 and 1:9. The chamber initially pumped to 5×10^{-6} Torr base pressure and the RF power 60 W.	Nb-doped $SrTiO_3$ (001); x = 20–25 nm d = 200 nm
Soft Chemical method and **Spin**-coating [31]	**$BiFeO_3$**; Hours; Iron(III) nitrate nonahydrate, Bismuth nitrate, ethylene glycol, citric acid	The precursor solutions of bismuth and iron were prepared by adding to the ethylene glycol and citric acid. The molar ratio of metal/citric acid/ ethylene glycol was 1/4/16. The viscosity of the resulting solution adjusted to 20 cP by controlling the water content using a Brookfield viscosimeter. The films spin-coated from a BFO deposition solution onto Pt/Ti/SiO_2/Si substrates and annealed at 500°C/2 h.	Pt/Ti/SiO_2/Si; x = 53–58 nm d = 200–300 nm
Three in one solution approach [32]	**$BiFeO_3$**; Days; $Bi(NO_3)_3 \cdot 5H_2O$, $Fe(NO_3)_3 \cdot 9H_2O$, $CH_3N(CH_2CH_2OH)_2$, $HO(CH_2)_3OH$, CH_3CH_2OH,	Metal nitrates first dissolved at room temperature in a mixture of MDEA and 1,3-propanediol solvents using a molar ratio of 1:5:10 (metal:MDEA: diol) and the complexing reactions promoted by heating at 150–190°C/ 120 min. After diluting with ethanol, stable solutions of Bi-MDEA and Fe-MDEA precursors obtained with a concentration of 0.3 M. Stoichiometric amounts of these solutions mixed together to get $BiFeO_3$ (0.3 M). For $BiFeO_3$ film, layers from the respective precursor deposited on Pt/TiO_2/SiO_2/(100)Si substrates by spin coating (2000 rpm/45 s) and dried on a hot	Flexible polymer substrates (polyimide); x = 55 nm; d = 185 nm

Synthesis method	Chemical composition; reaction time; precursor salts	Brief synthesis procedure & heating conditions of processing	Substrate; grain size (x); film thickness (d)
		plate at 150°C/5 min, and subjected to UV irradiation in oxygen at 150°C/30 min.	
Wet Chemical route [33]	**$BiFeO_3$**; Hours; $Bi(NO_3)_3$. $5H_2O$, $Fe(NO_3)_3$. $9H_2O$, ammonium hydroxide, ethanol, acetic acid, diethanolamine, 2-ethoxyethanol	In the first step, an aqueous solution of ferric nitrate treated with ammonium hydroxide (25%) solution and the precipitate washed with distilled water. 4 mL of acetic acid (95%) then added into it under stirring and heated at 75°C/30 min, and five drops (0.00419 M) of diethanolamine (DEA) added. In the second step, a required amount of bismuth nitrate dissolved in 20 mL 2-ethoxyethanol solvent and then 0.5 mL acetic acid (95%) also added into it and also added five drops (0.00419 M) of DEA. Both the solution mixed together and the precursor deposited by dip coating (speed 6 cm/min) and spin coating process (4000 rpm for 20 s) on ITO substrates.	ITO; $\boldsymbol{x}$ = 5 nm; $\boldsymbol{d}$ = ~ 2 μm and ~ 200 nm
Laser molecular-beam epitaxy [34]	**$BiFeO_3$**; Hours	Epitaxial BFO thin films of various thicknesses of 2, 4, 8, 19, 38, 57, 94, 141, 188, 283, and 449 nm were deposited on (001)-oriented STO and slightly Nb-doped STO single crystals by a laser molecular-beam epitaxy system. The films deposited at 560°C using an excimer XeCl laser (1.5 J/cm^2, 308 nm, 2 Hz) at oxygen pressure of 2 Pa (2×10^{-2} mbar). After deposition, the films subjected into situ annealed for 30 min then cooled down slowly to room temperature [35].	(001)-oriented STO; $\boldsymbol{x}$ = 8 nm; $\boldsymbol{d}$ = 8–449 nm

Bold emphasis has more significance.

Table 1.
Synthesis methods used to prepare multiferroic $BiFeO_3$ (BFO) thin films.

induced magnetization switching observed. It is due to the effect of large heteroepitaxial interface and reduced clamping effect. In **Figure 4a**, the new structure that fully epitaxial multilayered nanodot array, *i.e.* 0–0 composite which combines the advantages of 2–2 and 1–3 geometries to obtain a better understanding of extrinsic ME coupling and build prototypes for high density multistate memory devices. In 2–2 type, the horizontal stacking, *i.e.* an epitaxial multilayer or superlattice can provide more flexibility for material design, composition control, and layer arrangement.

2.2.2 SEM images of $BaTiO_3$-$CoFe_2O_4$ (BTO-CFO) two-layered nanodots

A typical scanning electron microscope (SEM) image of the as obtained STO/AAO substrate is given in **Figure 4b** [39]. There is no gap found between STO and

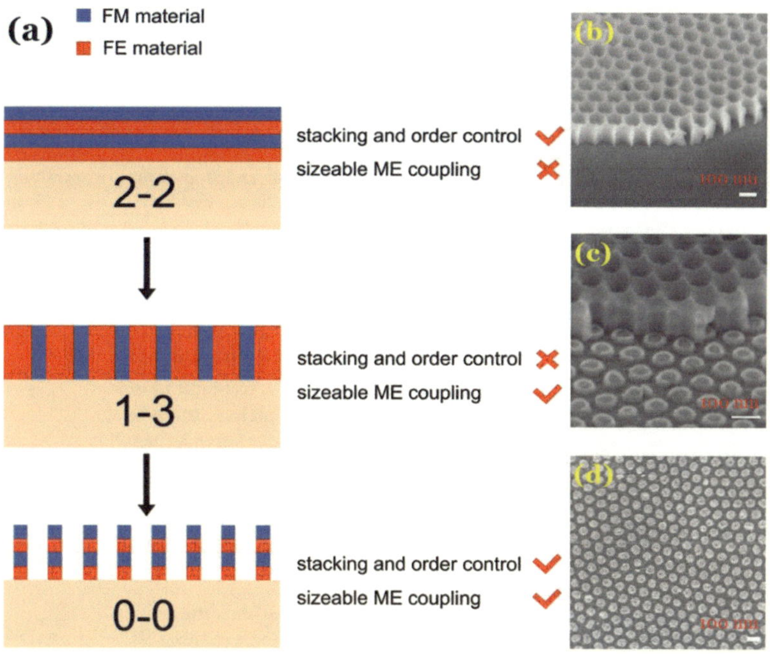

Figure 4.
(a) The film-on-substrate geometry in ME composites. SEM images of (b) as-transferred anodic aluminum oxide (AAO) mask on STO substrate, (c) BTO nanodots with partly removed AAO in first layer deposition, (d) BTO/CFO two-layered nanodot arrays [39].

the AAO membrane (thickness ~ 120 nm and pore diameters ~65 nm). The thin films are synthesized by PLD process. In **Figure 4c** and **d**, the highly ordered nanodot arrays with flat surfaces and diameters around 65 nm were obtained. The AAO was fabricated through a self-assembly process, the periodic area is only μm^2 in size.

2.2.3 Synthesis techniques for $BaTiO_3$ thin films

The thin film processing issues associated with various techniques such as the formation of side phases and the difficulty in controlling stoichiometry, film thickness, and sample crystallinity. For this we have summarized some synthesis method for $BaTiO_3$ based multiferroic thin films described in **Table 2**.

2.3 $PbTiO_3$ multiferroics

Since ferroelectricity in perovskites ABO_3 commonly involves B-site transition-metal ions with a formal electron configuration d^0 (e.g. Ti^{4+}, Nb^{5+}, etc.), while magnetism requires TM cations with partially filled d states. A lot of studies on multiferroics concentrate on the Bi-based perovskites, *i.e.* $BiFeO_3$ and $BiMnO_3$, where the ferroelectricity mainly arises from the lone pair of 6 s electrons [53]. But the basic physics in the ferroelectric thin films is similar to the bulk state in addition to some specific properties in thin-film like interface strain and stress, dead layer effect etc. [54]. In perovskites, $PbTiO_3$ also have multiferroic behavior with TM doping or composites substitution and have spontaneous polarization parallel to the *c* axis. The lattice structure of $PbTiO_3$ is tetragonal below the T_C and turns into cubic and paraelectric above T_C. The value of T_C for bulk $PbTiO_3$ lies in 490–493°C which depends upon synthesis process, size or defect effect. The displacement of the Ti

Synthesis method	Chemical composition; reaction time; precursor salts	Brief synthesis procedure & heating conditions of processing	Substrate; grain size (x); film thickness (d)
Atomic oxygen assisted molecular beam epitaxy (**AO-MBE**) [40]	**$BaTiO_3/CoFe_2O_4$**; Hours; Metallic individual Ba, Ti, Co, Fe	Films deposited by evaporation of metallic individual Ba, Ti, Co, Fe from dedicated Knudsen cells assisted by a high brilliance (350 W- 10^{-7} mbar oxygen -baratron 3.1 tr) atomic oxygen plasma with growth rate ~ 0.5-2 Å/min. Metal Ba kept in oil before insertion in the chamber because of its high reactivity with water and the oil removed using C_6H_{12} ultrasound assisted baths. The wet Ba metal introduced in MBE system just prior pumping, the procedure allows easy and fast outgassing of the Ba charge. This approach has advantage of enabling independent Ti, Ba, Co, Fe dosing as well as oxygen dosing.	Nb:$SrTiO_3$ (001) and Pt (001); $x < 25$ nm; $d \sim 30$ nm
Chemical bath deposition [41]	**$BaTiO_3$**; Hours; $Ba(CH_3COO)_2$, Ti $(C_4H_9O)_,$ acetic acid, ethylene glycol, butanol	In chemical bath deposition, the substrate is immersed in a solution containing the precursors which depends upon parameters like bath temperature, pH of the solution, the molarity of concentration, and time. Two solutions of barium acetate in butanol and acetic acid, and tetra-butyl titanate in butanol and ethylene glycol are mixed to get a final product. Thin films of $BaTiO_3$ grown on glass substrates with solution temperature 78° C and annealed at 300°C/1 h.	Glass; x = 23–30 nm
Chemical solution deposition [42]	**$BaTiO_3$-$CoFe_2O_4$**; Hours; $Ba(CH_3COO)_2$, $(C_4H_9O)_4Ti$, Fe $(NO_3)_3.9H_2O$, Co $(CH_3COO)_2$. $24H_2O$, 2-methoxyethanol, acetic acid, ethanolamine	2-Methoxyethanol and acetic acid used as the solvents with a ratio of 3:2 in volume for preparing both BTO and CFO precursors. Ethanolamine added to adjust the viscosity. The films spin-coated on Pt/Ti/SiO_2/Si(100) substrate with 6000 rpm/10s. The BTO precursor deposited on the substrate firstly, and baked in a 350°C preheated tube furnace for 10 min in air after each coating and pre-sintered for 10 min/700°C. After that the CFO layer deposited on the BTO layer, finally, the film annealed for 700° C/2 h.	Pt/Ti/SiO_2/Si (100); $x \sim 11$ nm; d = 500 m
Solid state reaction and DC sputtering [43]	**$La_{0.7}Ca_{0.3}MnO_3$/ $BaTiO_3$**; Days; Oxides of La, Ca, Mn, Ba, Ti	The base material is prepared by a solid state or any chemical synthesis method. Films were grown on [001] oriented BTO and [100] oriented STO single crystal substrates at high O_2 pressure dc sputtering at 900°C.	$BaTiO_3$; x = 300 nm
Electrophoretic Deposition [44]	**$BaTiO_3/CoFe_2O_4$**; Days; Acetates of Ba, Fe, Co, $Ti(C_4H_9)O_4$, ethanol, acetic acid, NaOH,	$BaTiO_3$ and $CoFe_2O_4$ powders were prepared using **sol–gel** and **chemical co-precipitation** method, respectively. Tween 80 and polyethylene imine polymers used as dispersants, and polytetra fluoro ethylene made	Al_2O_3/Pt; $x \sim 100$ nm; $d \sim 50$ μm

Synthesis method	Chemical composition; reaction time; precursor salts	Brief synthesis procedure & heating conditions of processing	Substrate; grain size (x); film thickness (d)
	Tween 80, polyethylene imine	rectangular cell used as the electrophoresis tank for the electrophoretic deposition where Al_2O_3/Pt acted as cathode for film deposition. The deposited films sintered at 600 and 1200°C	
Liquid phase deposition [45]	**5–50 wt % $CoFe_2O_4$-$BaTiO_3$**; Day; Barium, titanium(IV) isopropoxide, cobalt and iron acetylacetonate, anhydrous benzyl alcohol, acetone, 2-propanol, methoxyethoxy) ethoxy] acetic acid, and hexane, ethanol	For the synthesis of $BaTiO_3$ nanoparticles, 137 mg of dendritic Ba dissolved in 5 mL of degassed acetophenone at 80°C in an argon filled glovebox, with dropwise addition of a titanium isopropoxide and the mixture transferred to a microwave vial, sealed, and exposed to microwave irradiation at 220°C/30 min. For $CoFe_2O_4$ synthesis, 353 mg of $Fe(C_5H_7O_2)_3$ and 120 mg of $Co(C_5H_7O_2)_2$ mixed with 5 mL of benzyl alcohol in an argon-filled glovebox and subjected to microwave irradiation at 180°C/30 min in a 10 mL Teflon-capped glass vessel. The $BaTiO_3$ and $CoFe_2O_4$ nanoparticles are centrifuged off at 4000 rpm, washed with ethanol, and finally sonicated for 45 min in 0.3 M ethanolic MEEAA solution using a ultrasonic cleaner. In the next step, an excessive amount of hexane (5:1 hexane to ethanol ratio) is added, and the mixture is centrifuged to separate the nanoparticles. Afterward, the $BaTiO_3$ and $CoFe_2O_4$ nanoparticles redispersed in ethanol with sonication. The co-dispersions $BaTiO_3$/$CoFe_2O_4$ mixure deposited using spin-coating (1000 rpms/20s) and heated at 500°C and film annealed at 700°C/2 h.	$Pt/TiO_2/SiO_2/Si$; x = 13 nm; d = 400–500 nm
Sol–gel followed magnetron sputtering [13, 46]	**$BaTiO_3$-$Ni_{0.5}Zn_{0.5}Fe_2O_4$**; Hours	The BTO and NZFO synthesized via sol gel and citric acid combustion method and sintered at 1200°C/2 h. The sputtering atmosphere was set at 0.4 O_2 and 0.6 Ar_2 with a total pressure of 0.6×10^{-3} bar, and the sputtering power adjusted between 80 W and 250 W (sputtering time selected between 2 and 6 h). The deposited films annealed at 600 -820°C.	(100) and (111) Si; $x \sim 20$ nm; $d \sim 100$ nm
Metallo-organic decompositi (**MOD**) [38]	**$MFe_2O4/BaTiO_3$**; Days; xylene, 2-ethylhexanoic acid, KOH, Ba, Fe, Mn, Co, Ni, Zn nitrates, $C_{16}H_{36}O_4Ti$, polyethylene glycol, acetone, isopropanol	0.1 M solution of 2-ethylhexanoic acid was neutralized with KOH and then mixed with respective nitrates of Ba, Mn,Co, Ni and Zn. Each film was taken in 1:1 ratio. Final solution was refluxed at 110°C/10 h and then PEG used as binder in the coating MFO/BFO solution. The solution coated on $Pt/Ti/SiO_2/Si$ substrate using spin-coating (4500 rpm/60 s) and annealed at 600°C/3 h.	$Pt/Ti/SiO_2/Si$; x = 20–137 nm; $d \sim 600$ nm

Synthesis method	Chemical composition; reaction time; precursor salts	Brief synthesis procedure & heating conditions of processing	Substrate; grain size (x); film thickness (d)
Off-axis RF magnetron sputtering [47]	**$BaTiO_3$-$BiFeO_3$**; Hours	Buffered oxide etchant (BOE) used to treat the surface of STO substrate followed by annealing at 900°C in an O_2 atmosphere. LSMO thin film deposited on STO substrate at 600°C at 5 m Torr with 8:2 O_2:Ar gas ratio and 4.93 W cm^{-2} of RF power. $BaTiO_3$-$BiFeO_3$ thin film grown on the epitaxial LSMO thin film at 400°C via in-situ deposition under same RF power and pressure with 1:9 O_2:Ar gas ratio.	$SrTiO_3$; x = 200–500 nm; d = 80–500 nm
Polymer-assisted deposition [48]	**$BaTiO_3$-$CoFe_2O_4$**; Hours; EDTA, PEI, Ba, Fe, Co, Ti chlorides,	Barium chloride or iron chloride added to the solution of an ethylenediammine tetraacetic acid (EDTA) and Polyethylenimine (PEI). Cobalt chloride added to PEI solution at pH -7, and the mixture of titanium chloride and peroxide added to the mixture of PEI and EDTA to maintain pH 7.5. All the solutions mixed with a molar ratio of Ba: Ti:Co:Fe = 1:1:1:2 followed by spin coating on LAO substrates at 2000 rpm/ 30s and the samples heated at 500°C and annealed at 900°C/1 h.	$LaAlO_3$ (100); $x \sim 90$ nm; $d \sim 100$ nm
One-step Pulsed laser deposition followed Solid State Reaction [49]	**$(BaTiO_3)_{0.8}$: $(La_{0.7}Sr_{0.3}MnO_3)_{0.2}$**; Days; $BaTiO_3$, La_2O_3, $SrCO_3$, MnO_2	Solid state sintering method used to synthesize target sample BTO:LSMO with atomic ratio 8:2. The STO (001) substrates used to deposit thin films via pulsed laser deposition. The temperature of 750°C in a 200 mTorr oxygen atmosphere was maintained during deposition and a 400 mJ laser was used to shot the targets for 3000 pulses. The substrate distances kept at 4.5 cm, and cooling of the films in 200 mTorr oxygen atmosphere to room temperature was done after deposition.	STO (001) x = nanopillars; d = 90–117 nm
Laser molecular beam epitaxy [50]	**$(Bi_{0.5}Na_{0.5})TiO_3$-$BaTiO_3$ (Mn** doping); Hours; Oxides of Bi, Na, Ti, Ba, Mn	Conventional sintering method used to synthesize Mn doped BNBTMn by using excess of Bi_2O_3 thin layer of $La_{2/3}Sr_{1/3}MnO_3$ deposited on STO substrate at 800°C. The films grown on LSMO coated STO substrate at 750°C by deposition of LSMO dot top electrodes through a shadow mask on the surface of BNBTMn at 700°C. The laser with power of 5 J/cm^2 and oxygen partial pressure of 200 mTorr was used for all depositions.	(001)STO; $d \sim 400$ nm
Two-step anodization process followed spin-coating [51]	**$BaTiO_3$-$CoFe_2O_3$**; Hours; Ti foil, NH_4F, ethylene glycol, dimethyl sulfoxide, $Ba(OH)_2$, $Co(CH_3COO)_2 \cdot 4H_2O$, $Fe(NO_3)_3 \cdot 9H_2O$, 2-methoxyethanol	Two step anodization process was used to prepare TNTs. First step included anodization of Ti foil in an ethylene glycol-based electrolyte containing 0.09 M NH_4F and 1.24 M H_2O at 60 V. Second step included the use of mixture of ethylene glycol and dimethyl sulfoxide containing 0.1 M NH_4F and	Ti substrate; $x \sim$ nanopillars (diameter 40–50 nm, length 2600 nm)

Synthesis method	Chemical composition; reaction time; precursor salts	Brief synthesis procedure & heating conditions of processing	Substrate; grain size (x); film thickness (d)
		1.5 M H_2O at 60 V, 60°C/min followed by 15 minute treatment with electrolyte. TNTs were then treated with $Ba(OH)_2$ aqueous solution in a Teflon lined hydrothermal reactor at 180°C/2 h. CFO precursors were prepared by mixing Co $(CH_3COO)_2.4H_2O$ and $Fe(NO_3).9H_2O$ in 2-methoxy ethanol and H_2O followed by ultrasonication for 1 h. The spin coating of BTO with CFO was done to prepare BTO/CFO nanocomposite at 4000 rpm and 260°C/10 min. Final sample calcined at 600°C/3 h	
Tape casting method [52]	**Fe_3O_4@$BaTiO_3$/P (VDF-HFP)**; Days; $FeCl_3$, CH_3COOK, EG, $C_4H_{10}O$, $C_{19}H_{42}BrN$, $C_{16}H_{36}O_4Ti$, $Ba(OH)_2$, CH_3COOH, N, N-C_3H_7NO, P(VDF-HFP)	The FO submicron spheres prepared by a solvothermal synthesis and the FO@BTO core-shell particles prepared by a two-step hydrolysis-hydrothermal method. The required amount of core-shell FO@BTO powder dispersed in 10 ml DMF and placed in an ultrasound bath for 4 h. Then a calculated amount of P(VDF-HFP) dissolved into the suspension using an ultrasonic bath with stirring. The films of P(VDF-HFP) and core-shell particles prepared by tape casting and dried at 80°C and heated at 200°C/7 min and quenched in ice water immediately.	x = 352 nm (FO), 8 nm (BTO); d = 15 μm

Bold emphasis has more significance.

Table 2.
Synthesis method used to process $BaTiO_3$ based multiferroic thin films.

atom in the O octahedron gives the spontaneous polarization of the $PbTiO_3$ due to the vector of polarization from the center of the O octahedron to the Ti atom. Through the substitutions for Pb or Ti or both cations, the ferroelectric properties could be widely modified and the structures could show various characters. Due to scarcity of multiferroics, $PbTiO_3$ is being extensively studied for induction of magnetism. The large 'Pb' cation is coordinated by twelve oxygens in a cub-octahedral way, while the smaller 'Ti' cation is octahedrally coordinated with six oxygens [55]. The large degree of ferroelectricity in $PbTiO_3$ occurs to the contribution of covalent bonds between both Pb 6p - O 2p and Ti 3d - O 2p states. The covalent bond between Pb 6p and O 2p lowers the non-centrosymmetric symmetry towards the ferroelectric state by weakening the ionic core repulsion between Pb and the nearest oxygen ions.

2.3.1 ME coupling due to ferroelectric polarization under an external magnetic field

Figure 5a shows spontaneous polarization (P-E loop) under the influence of a magnetic field (H = 0–0.5 T) for $Pb_{0.7}Sr_{0.3}(Fe_{0.012}Ti_{0.988})O_3$ thin film. This thin film prepared by a MOD method and deposited on $Pt/Ti/SiO_2/Si$ substrate using spin-coating [56]. The magnetic switching from $+P_r$ to zero at 0.5 T have been observed.

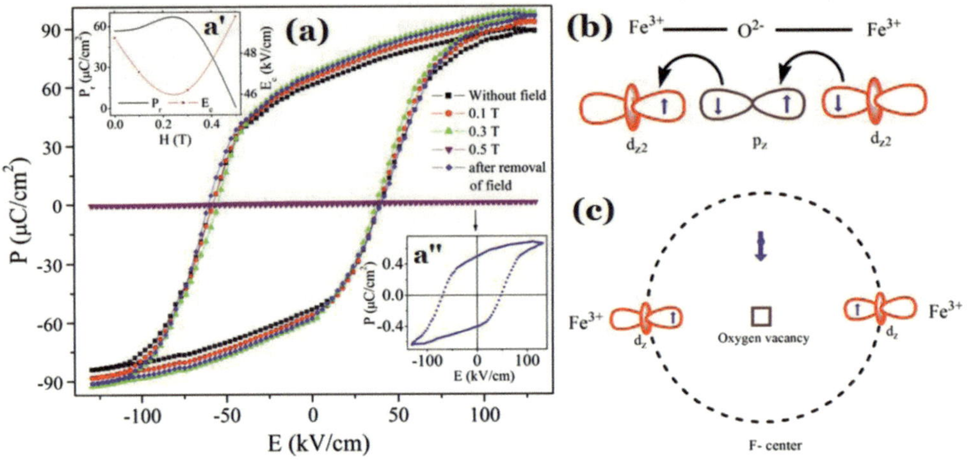

Figure 5.
(a) Ferroelectric hysteresis under an external magnetic field (0–0.5 T) for $Pb_{0.7}Sr_{0.3}(Fe_{0.012}Ti_{0.988})O_3$ thin film. (a′) Variation of P_r and E_c with applied H. (a) P ~ 0 at 0.5 T (contribution by electrode) [56]. (b and c) F-center exchange mechanism responsible for room-temperature ferromagnetism in Fe-doped $PbTiO_3$ [57].

In **Figure 5a′**, the polarization is continuously increased with increasing H up to 0.3 T which is an indication for the coupling between the polarization and magnetization. When a magnetic field applied to a ME material, the material is strained to induce a stress on the piezoelectrics to generates the electric field. This field related with ferroelectric domains resulting into an increase in polarization. **Figure 5a″** shows a low polarization response with lossy hysteresis at 0.5 T, which may have been caused by the electrode.

2.3.2 Mechanism for ferromagnetism in Fe-doped $PbTiO_3$

F-center exchange (FCE) mechanism is used to evaluate the magnetic origin in $Pb(Ti_{1-x}Fe_x)O_3$ thin films as shown in **Figure 5b** and **c** [54, 57]. As the tetravalent Ti^{4+} is replaced by trivalent Fe^{3+} cations, there is generation of oxygen vacancies to cause charge neutrality in the multiferroic. An electron trapped in the oxygen vacancy might to produce an F center, where the electron occupies an orbital to overlaps the *d* shells of both iron neighbors. As the unoccupied spin orbitals are minority in the $3d^5$ of Fe^{3+}, the trapped electron and the iron neighbors have opposite spin direction. Therefore, the system becomes ferromagnetic due to super-exchange effect generates antiferromagnetism.

2.3.3 Synthesis techniques for $PbTiO_3$ thin films

To the investigation of bulk ferroelectric materials, the compositional adjustment has plays a major role to study new ferroelectric. However for thin films, the suitability of synthesis method for processing of epitaxial ferroelectric plays key feature to change the polarization behavior due to some technical features related with film geometry on substrate. In **Table 3**, we have introduced some typical synthesis routes for $PbTiO_3$-based multiferroic thin films fabrication.

2.4 Multiferroic $CaTiO_3$ systems

Since $CaTiO_3$ (CTO) is prototype for the perovskite structure [67]. At room temperature, CTO has the orthorhombic $GdFeO_3$ type crystal structure (a = 5.38 Å,

Synthesis method	Chemical composition; reaction time; precursor salts	Brief synthesis procedure & heating conditions of processing	Substrate; grain size (x); film thickness (d)
Chemical solution deposition [58]	**$0.7BiFeO_3$–$0.3PbTiO_3$ (Mn doping); Days;** $Bi(O^tC_5H_{11})_3$, Fe $(OC_2H_5)_3$, Pb $(CH_3COO)_2$, Ti $(O^iC_3H_7)_4$, Mn $(O^iC_3H_7)_2$, 2-Methoxyethanol	Calculated amounts of $Bi(O^tC_5H_{11})_3$, Fe $(OC_2H_5)_3$, $Pb(CH_3COO)_2$, $Ti(O^iC_3H_7)_4$ and $Mn(O^iC_3H_7)_2$ were dissolved in 2-Methoxyethanol which acted as solvent. The Mn content set at 5 mol% for Fe site, solution refluxed for 20 h, and the entire process conducted in dry N_2 environment. Thin films were fabricated via spin coating at 2500 rpm/30s on $Pt/TiO_x/SiO_2/Si$ substrates followed by drying at 150°C/5 min and calcined at 400°C/1 h in O_2 flow.	$Pt/TiO_x/SiO_2/Si$; $x \sim 60$ nm; $d \sim 500$ nm
DC magnetron sputtering [59]	**$(Co/Ni)_4/Pb(Mg_{1/3}Nb_{2/3})O_3$-$PbTiO_3$;** Days	Multi-layered multiferroic films grown on a (001)-cut single crystal PMN-PT substrate at pressure below 7×10^{-6} (5×10^{-8} Torr), and Ar^+ ion bombarded to remove organic contamination. The various multilayered films with different thickness where 3 nm Ta adhesion layer was followed by 2 nm Pt to encourage (111) face-centered-cubic growth of subsequent film. Ni thickness was altered between 0.15 and 0.9 nm to synthesize five Co/Ni multi-layered films PMN-Pt: [Co (0.6 nm)/Ni(tNi]$_{x4}$/Co (0.6). The capping layers were identical initiating with a 2 nm Pt layer with a protective 3 nm Ta capping electrode. A fixed power of 40 W and a pressure of 0.7 Pa are used.	[001]-cut PMN-PT; d = ultrathin
Hydrothermal method [60]	**$LaFeO_3$-$PbTiO_3$;** Days; $Pb(NO_3)_2$, La $(NO_3)_3.6H_2O$, Fe $(NO_3)_3.9H_2O$, TiO_2, KOH	$Pb(NO_3)_2$, $La(NO_3)_3.6H_2O$, Fe $(NO_3)_3.9H_2O$ and TiO_2 were thoroughly mixed in 9 M solution of KOH, and mixture stirred for 2 h and then transferred to Teflon-lined stainless autoclave having a single crystal (001) oriented NSTO at the bottom of the autoclave. The sealed autoclave heated at 200°C/36 h	**Nb-$SrTiO_3$ (100);** $x \sim 20$ nm; $d \sim 250$ nm
Laser ablation [61]	**$PbTiO_3$-$CoFe_2O_4$;** Hours	Pure PTO and CFO target were prepared by any of chemical synthesis method. $PbTiO_3$-$CoFe_2O_4$ thin films formed by depositing $CoFe_2O_4$ and $PbTiO_3$ via laser ablation in the superlattice spread. The average composition changes continuously from one end (pure PTO) to the other (pure CFO) films deposited on MgO substrate at 600°C and oxygen partial pressure of 65 mTorr. The energy used for ablation was ~2 J/cm^2. The total thickness at each position on the spread was 300 nm, and sample ~ 6 mm long in the direction of spread.	(100) MgO; $x \sim 30$ nm; $d \sim 12.6$
Metal–organic chemical vapor	**$CoFe_2O_4$-$PbTiO_3$;** Hours; Co tris(2,2,6,6-	The c-axis orientation chosen for substrate, and the procedure included evaporation of metal organic precursors	(001) $SrTiO_3$; $x \sim 20$ nm; $d \sim 280$ nm

Synthesis method	Chemical composition; reaction time; precursor salts	Brief synthesis procedure & heating conditions of processing	Substrate; grain size (x); film thickness (d)
deposition [62]	tetramethyl-3,5-heptanedionate), Fe(III) tris(2,2,6,6-tetramethyl-3,5-heptanedionate), tetraethyl Pb, Ti(IV) t-butoixde, O_2 gas	from Co tris(2,2,6,6-tetramethyl-3,5-heptanedionate), Fe(III) tris(2,2,6,6-tetramethyl-3,5-heptanedionate), tetraethyl lead, titanium (IV) t-butoixde and O_2 gas followed by entering of evaporated precursors into the deposition chamber of MOVCD system. The O_2 flow increased for completion of reactions. The cooling water used to avoid deposition on the walls and to create temperature gradient, and the nanocomposites are kept at room temperature under O_2 gas flow to ensure cooling.	
Off-axis magnetron sputtering and sol–gel spin-coating [63]	NiFe/Pb($Mg_{1/3}Nb_{2/3}$)O_3-$PbTiO_3$; Hours	Sol–gel spin coating method used first to deposit PMN-PT films on Pt/Ti/SiO_2/Si substrates, and precursors spin coated at 4000 rpm/30s followed by pyrolyzation at 450°C/5 min. This process was repeated until desired thickness is obtained followed by annealing at 650°C/10 min in air. The sputtering of NiFe thin films on PMN-PT films via off-axis magnetron sputtering at 200°C was the second step. The base pressure of 3×10^{-7} Torr maintained before deposition. The deposition of NiFe thin films was done by applying power of 60 W and Ar gas pressure of 2×10^{-7} Torr.	Pt(111)/Ti/SiO_2/Si; $x \sim$ 10–250 nm; $d \sim$ 30–220 nm
Pulsed laser deposition [64]	**$CoFe_2O_4$-$PbTiO_3$**; Days	Composite targets with preselected fixed compositions used to prepare $CoFe_2O_4$-$PbTiO_3$ thin films. The (001), (110) and (111) $SrTiO_3$ acted as substrate. The 630°C temperature maintained for the substrate and oxygen pressure of 100 mTorr was maintained in the deposition chamber. The approximately 200–230 nm thick films deposited via laser density of 1.6 Jcm^{-2} at a growth rate of 2–3 nm min^{-1}	$SrTiO_3$; x = 50 100 nm; $d \sim$ 200–230 nm
Sol–gel followed by dip coating [65]	**$PbTiO_3$/$NiFe_2O_4$**; Hours; Pb($CH_3COO)_2$, Ti ($OC_4H_9)_4$, Ni ($CH_3COO)_2$, Fe($NO_3)_3$, CH_3COOH, $C_3H_8O_2$	The composition of the raw materials was kept in the mole ratio of Pb:Ti:Ni:Fe of 1:1:1:2 for the final thin film of $PbTiO_3$ and $NiFe_2O_4$. The thin films prepared on silicon substrates using the sol precursor by dip coating, followed by a heat-treatment at 550°C/8 min plus a post heat-treatment at temperatures of 550-950°C/1 h. The acetic acid (CH_3COOH) and ethylene glycol monomethyl ether ($CH_3OCH_2CH_2OH$) acting as solvents during precursor preparation.	Si; x = 80–170 nm; $d \sim$ 1.2 µm
MOD-spin coating [66]	**$Pb_{0.7}Sr_{0.3}(Fe_{0.012}Ti_{0.988})O_3$**; Days;	The precursors solution of $Pb_{0.7}Sr_{0.3}(Fe_{0.012}Ti_{0.988})O_3$ thin film prepared using Pb 2-ethylhexanoate	Pt/Ti/SiO_2/Si; x = 16 nm; d = 500 nm

Synthesis method	Chemical composition; reaction time; precursor salts	Brief synthesis procedure & heating conditions of processing	Substrate; grain size (*x*); film thickness (*d*)
	xylene, 2-ethylhexanoic acid, KOH, Pb, Fe, Sr. nitrates, $C_{16}H_{36}O_4Ti$, polyethylene glycol, acetone, isopropanol	$(C_7H_{15}COO)_2Pb$, Sr. 2-ethylhexanoate $(C_7H_{15}COO)_2Sr$, Fe 3-ethylhexanoate and tetra-n-butyl orthotitanate in xylene. An approximately 20 mol% excess of Pb precursor added to compensate for the loss of Pb during heating. The solution refluxed at 110°C with constant stirring for 10 h for homogeneous mixing. The solution spin coated on $Pt/Ti/SiO_2/Si$ substrates by the spin coating technique with 4300 rpm/60s dried at 350°C/5 min to remove the solvent and organic residuals. The final film annealed at 650°C/3 h.	

Bold emphasis has more significance.

Table 3.
Synthesis method used to prepare multiferroic $PbTiO_3$ based thin films.

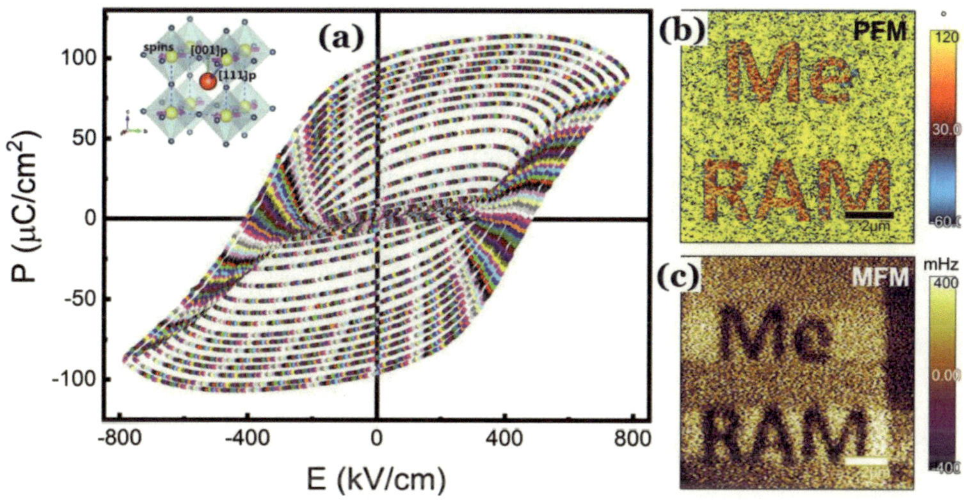

Figure 6.
(a) Spontaneous polarization in $0.85BiTi_{0.1}Fe_{0.8}Mg_{0.1}O_3$–$0.15CaTiO_3$ (BTFM-CTO) thin film. (b) PFM image after electric writing by 10 V, and corresponding. (c) MFM image without an external magnetic field [69].

b = 5.44 Å, *c* = 7.65 Å) and exhibits cubic ($Pm\bar{3}m$) one above ~1575 K. Below this temperature, CTO has tetragonal (I4/mcm) phase for which transitions of orthorhombic (Pnmb) phase exist below ~1525 K [67]. This orthorhombic structure could create an epitaxial strain in thin films due to changing symmetry at the phase transitions by tilting TiO_6 oxygen octahedral away from cubic one. CTO has possessed its nonpolarity to antiferrodistortive (AFD) distortions of the TiO_6 octahedral rotations and cation displacements that resulting into an orthorhombic (Pbnm) symmetry and $a^-\ a^-\ c^+$ rotations in glazer notation [68]. The ferroelectric polarizations and transition temperatures are significantly enhanced in classical ferroelectric perovskite like $BaTiO_3$ and $PbTiO_3$, and the enhancement in lattice strain might to turn ferroelectricity. Otherwise nonpolar ferroelectrics such as $SrTiO_3$ and

$CaTiO_3$ reported strain-induced ferroelectric transition temperatures only at room temperature or below. It is also reported that the simultaneous doping from Ca and Ti ions into $BiFeO_3$ results to decrease electric conductivity and stabilize the polar rhombohedral phase, and the magnetic properties are enhanced by protect oxygen stoichiometry with Fe in +3 oxidation state [68].

2.4.1 Ferroelectricity in $0.85BiTi_{0.1}Fe_{0.8}Mg_{0.1}O_3$-$0.15CaTiO_3$ thin film

Figure 6a shows the ferroelectric polarization (P) for $0.85BiTi_{0.1}Fe_{0.8}Mg_{0.1}O_3$–$0.15CaTiO_3$ thin film at room temperature [69]. This P-E hysteresis loop shows a

Synthesis method	**Chemical composition; reaction time; precursor salts**	**Brief synthesis procedure & heating conditions of processing**	**Substrate; grain size (*x*); film thickness (*d*)**
Pulsed laser deposition [69]	**$BiTi_{0.1}Fe_{0.8}Mg_{0.1}O_3$-$CaTiO_3$**; Days; Oxides of Bi, Ti, Fe, Mg, Ca	Conventional solid state reaction method used to synthesize the ceramic target of $0.85BiTi_{0.1}Fe_{0.8}Mg_{0.1}O_3$–$0.15CaTiO_3$. PLD deposition (with the laser source at 355 nm and a repetition rate of 10 Hz) was used to deposit BTFMO thin films on $Pt/TiO_2/SiO_2/Si$ substrate at ~500°C/ 30 min followed by an in-situ thermal annealing process for 10 min and cooled down to room temperature.	$Pt/TiO_2/SiO_2/$ Si; ***x*** = 40 nm; ***d*** ~ 330 nm
Sol–gel using spin coating [70]	**$BiFeO_3$-$CaTiO_3$**; Hours; $Bi(NO_3)_3.5H_2O$, Fe $(NO_3)_3.9H_2O$, Ca $(NO_3)_2.4H_2O$, $[CH_3(CH_2)_3O]_4Ti$	The $Bi(NO_3)_3.5H_2O$ and $Fe(NO_3)_3.9H_2O$ were dissolved in a mixture of 2-methoxy ethanol and ethylene glycol. The acetic anhydride added to the solution in order to acquire pure BFO precursor. An appropriate ratios of Ca $(NO_3)_2$. $4H_2O$ and $[CH_3(CH_2)_3O]_4Ti$ to obtain BFO-CTO precursor has been prepared. As-synthesized precursor spin coated on $Pt/Ti/SiO_2/Si$ substrates at 3000 rpm/30s and preheated at 350°C/ 10 min, and annealed at 550°C/15 min.	$Pt/Ti/SiO_2/Si$; ***x*** = 48–110 nm; ***d*** = 300 nm
Sol–gel based solution processing [71]	**$BiTi_{(1-y)/2}Fe_yMg_{(1-y)/2}O_3$–$CaTiO_3$**; Days; Nitrates of Bi, Fe, Mg, Ca $(CH_3COO)_2.H_2O$, $C_{16}H_{36}O_4Ti$, $C_6H_8O_7$, $CH_3OCH_2CH_2OH$ (2MOE), acetic acid and propionic anhydride	Sol–gel precursor solutions prepared by mixing nitrates of Bi, Fe, Mg, Ca $(CH_3COO)_2.H_2O$ and $C_{16}H_{36}O_4Ti$ with $C_6H_8O_7$ and $CH_3OCH_2CH_2OH$ under constant stirring which treated with 2MOE, acetic acid and propionic anhydride. This precursor solution deposited on $SrTiO_3/La_{0.7}Sr_{0.3}MnO_3$ substrate via sol–gel two-step solution processing method.	$SrTiO_3/$ $La_{0.7}Sr_{0.3}MnO_3$; ***d*** ~ 30.14 nm
Hydrothermal epitaxy [72]	**$CaTiO_3$**; Day; CaO, TiO_2, KOH	The CaO and TiO_2 were dissolved in a mixture of KOH and H_2O. The solution was put in Teflon-lined autoclave and the substrate put in the bottom of the autoclave, and heated at 200°C/12 h. As-synthesized samples were washed with deionized water and dried in air.	Nb:$SrTiO_3$ (001); *x* ~ 40 nm; ***d*** ~ 350 nm

Bold emphasis has more significance.

Table 4.
Synthesis methods used for fabrication of $CaTiO_3$ based multiferroic thin film.

large remanent polarization of $2Pr = 191.6\ \mu C\ cm^{-2}$ and coercive field (E_c) $2E_c = 681.8\ kV\ cm^{-1}$ without excluding the contribution from the leakage current.

2.4.2 ME coupling in $0.85BiTi_{0.1}Fe_{0.8}Mg_{0.1}O_3$-$0.15CaTiO_3$ thin film

The ME coupling might be studied by considering the magnetic field (H) effect on the ferroelectric polarization and magnetization (magnetic field on the domain switching) that shown in **Figure 6b** and **c** [69]. **Figure 6b** is the PFM image after electric lithography which is observed magnetic domain switching after the electric lithography (**Figure 5c**). Although, the magnetic domains are not completely switched, indicating that the ME coupling is stable at room temperature.

2.4.3 Synthesis techniques for $CaTiO_3$ thin films

Table 4 shown the chemical synthesis methods used in the preparation of multiferroic systems of $CaTiO_3$ thin films.

3. Conclusion and comparative features for synthesis methods used for $BiFeO_3$, $BaTiO_3$, $PbTiO_3$ and $CaTiO_3$ thin films

The chemical synthesis methods for thin film fabrication of perovskite $BiFeO_3$, $BaTiO_3$, $PbTiO_3$ and $CaTiO_3$ multiferroic systems and their composites have been described on the basis of their brief synthesis procedure, time of reaction, composition, heating conditions and morphology. Since thin film composite structures are mostly suitable for chip-device implementation. In various thin film processing methods of multiferroics, the vacuum based deposition like PLD may provide a large ME coupling, enabling the realization of complex microstructures including epitaxy, texture, or columnar distribution of magnetic nanopillars into a ferroelectric matrix. PLD enables the growth of high quality epitaxial thin films with substrate area for coated are small but to the making of uniform films on the wafer level is challenging.

The BFO films made via CSD needs richer oxygen environments for annealing and obtain crystalline films at relatively low temperatures. This CSD method has low cost, easily control of stoichiometry and easy to operate for large area films for a complex-shaped substrates. Among various CSD techniques, polymer assisted deposition (PAD) is most suitable to produce high quality epitaxial complex and multilayer-structured films (film thickness below 100 nm).

The multiferroic thin films deposited with RF sputtering performs high reproducibility with an accurate stoichiometry in a controlled process (but the film growth rate is low and the cost of fabrication is high), and considered to be the most suitable method of thin film preparation due to a smooth surface and dense structure that usable to fabricate the integrated circuit device.

Other chemical synthesis is sol–gel which has good reproducibility, low cost, thickness uniformity and moreover large area deposition and commonly used spin coating deposition. It simply prepared BFO thin films. Photosensitive sol–gel method can integrate the preparation with the fine patterning of the film which can simplify the lithography process remarkably.

The atomic layer deposition (ALD) provides additional degrees of freedom in design and fabrication of devices depending on domain wall optimization, adding the option of conformity in deposition of various geometries (it easily embedded nanoparticles, embedded nanopillars, and nanolaminates according to configurations, *i.e.* 0–3, 1–3, and 2–2) with unique thickness control and also to fabricate high-quality ultrathin films.

Another chemical technique is the chemical vapor deposition (CVD) which highly usable for BFO thin films deposition which has excellent substrate coverage, low-cost, ease of scale-up, control over thickness and morphology. Aerosol assisted CVD is not rely upon the use of highly volatile precursors, essential for typically high molecular weight heterometallic cluster compounds.

For shape control in multiferroics, hydrothermal method could provide an alternative approach to the synthesis of high-quality epitaxial BFO and BTO thin-film with low-cost. The hydrothermal reaction typically occurs at low temperature (<250°C) and high pressure (<15 MPa) originating from water vapor pressure. The spray pyrolysis method has been implemented for the production of porous and layer by layer films.

Acknowledgements

The author K.C. Verma thankfully acknowledges the financial support by UGC of Dr. DS Kothari Post Doctorate Fellowship [No. F4-2/2006(BSR)/PH/16-17/0066] and CSIR-HRDG for SRA (Pool Scientist) fellowship Grant No. B-12287 [SRA (Pool No): 9048-A].

Author details

Kuldeep Chand Verma[1,2*] and Manpreet Singh[3]

1 Materials Science and Sensor Applications (MSSA), CSIR-Central Scientific Instruments Organisation, Chandigarh, India

2 Department of Physics, Panjab University, Chandigarh, India

3 Department of Chemistry, Eternal University, Baru Sahib, Himachal Pradesh, India

*Address all correspondence to: dkuldeep.physics@gmail.com; kcv0309@gmail.com

References

[1] Ramesh R, Spaldin NA. Multiferroics: progress and prospects in thin films. Nat. Mater. 2007; **6**, 21-29.

[2] Verma KC, Goyal N, Kotnala RK. Tuning magnetism in $0.25BaTiO_3$-$0.75CoFe_2O_4$ hetero-nanostructure to control ferroelectric polarization. Physica B: Conden. Matter. 2019; **554**, 9-16.

[3] Wang J, Neaton JB, Zheng H, Vagarajan V, Ogale SB, Liu B, Viehland D, Vaithyanathan V, Schlom DG, Waghmare UV, paldinNA S, Rabe KM, Wuttig M, Ramesh R. Epitaxial $BiFeO_3$ Multiferroic Thin Film Heterostructures. Science 2003; **299**, 1719-1722.

[4] Chu YH, Martin LW, Holcomb MB, Gajek M, Han SJ, He Q *et al.* Electric-field control of local ferromagnetism using a magnetoelectric multiferroic. Nat. Mater. 2008; 7, 478-482.

[5] Kim DH, Ning S, Ross CA. Self-assembled multiferroic perovskite-spinel nanocomposite thin films: epitaxial growth, templating and integration on silicon. J. Mater. Chem. C 2019; 7, 9128-9148.

[6] Verma KC, Kotnala RK, Negi NS. Intrinsic study for magnetoelectric coupling in $Pb_{1-x}Sr_x(Fe_{0.012}Ti_{0.988})O_3$ Nanoparticles. Solid State Commun. 2009; **149**, 1743-1748.

[7] Verma KC, Kotnala RK. Tailoring the multiferroic behavior in $BiFeO_3$ nanostructures by Pb doping. RSC Adv. 2016; **6**, 57727-57738.

[8] Verma KC, Kotnala RK. Multiferroic magnetoelectric coupling and relaxor ferroelectric behavior in $0.7BiFeO_3$-$0.3BaTiO_3$ nanocrystals. Solid State Commun. 2011; **151**, 920-923.

[9] Yin L, Mi W. Progress in $BiFeO_3$-based heterostructures: materials, properties and applications. Nanoscale. 2020; **12**, 477.

[10] Verma KC, Synthesis and Characterization of Multiferroic $BiFeO_3$ for Data Storage. 2020; doi.org/10.5772/intechopen.94049, 106-135.

[11] Zhao T, Scholl A, Zavaliche F, Lee K, Barry M, Doran A, Cruz MP, Chu YH, Ederer C, Spaldin NA, Das RR, Kim DM *et al.* Electrical control of antiferromagnetic domains inmultiferroic $BiFeO_3$ films at room temperature. Nat. Mater. 2006; 5, 823-829.

[12] Bai F. *et al.* Destruction of spin cycloid in <111>$_c$-oriented $BiFeO_3$ thin films by epitaxial constraint: Enhanced polarization and release of latent magnetization. Appl. Phys. Lett. 2005; **86**, 032511.

[13] Gumiel C, Jardiel T, Calatayud DG, Vranken T, Van Bael MK, Hardy A *et al.* Nanostructure stabilization by low-temperature dopant pinning in multiferroic $BiFeO_3$-based thin films produced by aqueous chemical solution deposition. J. Mater. Chem. C. 2020; **8**, 4234-4245.

[14] Jalkanen P, Tuboltsev V, Marchand B, Savin A, Puttaswamy M, Vehkamaki M *et al.* Magnetic Properties of Polycrystalline Bismuth Ferrite Thin Films Grown by Atomic Layer Deposition. J. Phys. Chem. Lett. 2014; **5**, 4319-4323.

[15] Waghmare SD, Jadhav VV, Shaikh SF, Mane RS, Rhee JH, O'Dwyer C. Sprayed tungsten-doped and undoped bismuth ferrite nanostructured films for reducing and oxidizing gas sensor applications. Sens. Act. A. 2018; **271**, 37-43.

[16] Tomczyk M, Bretos I, Jimenez R, Mahajan A, Ramana EV, Calzada ML *et al.* Direct fabrication of $BiFeO_3$ thin

films on polyimide substrates for flexible electronics. **J. Mater. Chem.** C. 2017; **5**, 12529-12537.

[17] Moniz SJA, Cabrera RQ, Blackman CS, Tang J, Southern P, Weaverd PM *et al.* A simple, low-cost CVD route to thin films of $BiFeO_3$ for efficient water photo-oxidation. J. Mater. Chem. A. 2014; **2**, 2922-2927.

[18] Palgrave RG, Parkin IP. Aerosol Assisted Chemical Vapor Deposition Using Nanoparticle Precursors: A Route to Nanocomposite Thin Films. J. Am. Chem. Soc. 2006; **128**, 1587-1597.

[19] Tiwari D, Fermin DJ, Chaudhuri TK, Ray A, Solution Processed Bismuth Ferrite Thin Films for All-Oxide Solar Photovoltaics. J. Phys. Chem. C 2015; **119**, 5872-5877.

[20] Zargazia M, Entezari MH, A novel synthesis of forest like $BiFeO_3$ thin film: Photo-electrochemical studies and its application as a photocatalyst for phenol degradation. Appl. Surf. Sci. 2019; **483**, 793-802.

[21] Castro A, Martins MA, Ferreira LP, Godinho M, Vilarinho PM, Ferreira P, Multifunctional nanopatterned porous bismuth ferrite thin films. J. Mater. Chem. C. 2019; **7**, 7788-7797.

[22] Lee TK, Sung KD, Jung JH, Electric polarization and diode-like conduction in hydrothermally grown $BiFeO_3$ thin films, J. Alloys Compd. 2015; **622**, 734-737.

[23] Wang Z, Li Y, Viswan R, Hu B, Harris VG, Li J, Viehland D. Engineered Magnetic Shape Anisotropy in $BiFeO_3$-$CoFe_2O_4$ Self-Assembled Thin Films. ACS Nano. 2013; **7(4)**, 3447-3456.

[24] Han H, Lee JH, Jang HM. Low-Temperature Solid-State Synthesis of High-Purity $BiFeO_3$ Ceramic for Ferroic Thin-Film Deposition. Inorg. Chem. 2017; 56(**19**), 11911-11916.

[25] Azeem W, Riaz S, Bukhtiar A, Hussain SS, Xu Y, Naseem S. Ferromagnetic ordering and electromagnons in microwave synthesized $BiFeO_3$ thin films. J. Magn. Magn. Mater. 2019; **475**, 60-69.

[26] Islam MR, Zubair MA, Bashar MS, Rashid AKMB. $Bi_{0.9}Ho_{0.1}FeO_3/TiO_2$ Composite Thin Films: Synthesis and Study of Optical, Electrical and Magnetic Properties. Sci. Rep. 2019; **9**: 5205.

[27] Yan F, Zhao G, Song N, Zhao N, Chen Y. In situ synthesis and characterization of fine-patterned La and Mn co-doped $BiFeO_3$ film. J. Alloys Compd. 2013; **570**, 19-22.

[28] Vila-Fungueirino JM, Gomez A, Antoja-Lleonart J, Gazquez J, Magen C, Noheda B *et al.* Direct and Converse Piezoelectric Responses at the Nanoscale from Epitaxial $BiFeO_3$ Thin Films Grown by Polymer Assisted Deposition. **Nanoscale**. 2018; **10**, 20155-20161.

[29] Pham CD, Chang J, Zurbuchen MA, Chang JP. Synthesis and Characterization of $BiFeO_3$ Thin Films for Multiferroic Applications by Radical Enhanced Atomic Layer Deposition. Chem. Mater. 2015; 27, **21**, 7282-7288.

[30] Kim TC, Ojha S, Tian G, Lee SH, Jung HK, Choi JW *et al.* Self-assembled multiferroic epitaxial $BiFeO_3$-$CoFe_2O_4$ nanocomposite thin films grown by rf magnetron sputtering. J. **Mater. Chem. C**, 2018; **6**, 5552-5561.

[31] Simoes AZ, Riccardi CS, Santos MLD, Garcia FG, Longo E, Varela JA. Effect of annealing atmosphere on phase formation and electrical characteristics of bismuth ferrite thin films. Mater. Res. Bull. 2009; **44**, 1747-1752.

[32] Bretos I, Jimenez R, Ricote J, Sirera R, Calzada ML. Photoferroelectric Thin Films for Flexible Systems by a

Three-in-One Solution-Based Approach. Adv. Funct. Mater. 2020, **30(32)**, 2001897.

[33] Das S, Basu S, Mitra S, Chakravorty D, Mondal BN. Wet chemical route to transparent $BiFeO_3$ films on SiO_2 substrates. Thin Solid Films 2010; **518**, 4071-4075.

[34] Guo H, Zhao R, Jin K, Gu L, Xiao D, Yang Z *et al.* Interfacial-Strain-Induced Structural and Polarization Evolutions in Epitaxial Multiferroic $BiFeO_3$ (001) Thin Films. ACS Appl. Mater. Interf. 2015; 7, 2944-2951.

[35] Wang L, Wang Z, Jin KJ, Li JQ, Yang HX, Wang C *et al.* Effect of the Thickness of $BiFeO_3$ Layers on the Magnetic and Electric Properties of $BiFeO_3/La_{0.7}Sr_{0.3}MnO_3$ Heterostructures. Appl. Phys. Lett. 2013; **102**, 242902.

[36] Verma KC, Kotnala RK. Multiferroic approach for Cr,Mn,Fe,Co,Ni,Cu substituted $BaTiO_3$ Nanoparticles. Mater. Res. Express 2016; 3, 055006.

[37] Verma KC, Kotnala RK. Lattice Defects Induce Multiferroic Responses in Ce, La-Substituted $BaFe_{0.01}Ti_{0.99}O_3$ Nanostructures. J. Am. Ceram. Soc. 2016; 99[5], 1601-1608.

[38] Verma KC, Singh D, Kumar S, Kotnala RK. Multiferroic effects in $MFe_2O_4/BaTiO_3$ (M = Mn, Co, Ni, Zn) Nanocomposites. J. Alloys Compd. 2017; **709**, 344-355.

[39] Lu X, Kim Y, Goetze S, Li X, Dong S, Werner P *et al.* Magnetoelectric Coupling in Ordered Arrays of Multilayered Heteroepitaxial $BaTiO_3/CoFe_2O_4$ Nanodots. Nano Lett. 2011; **11**, 3202-3206.

[40] Stanescu D, Magnan H, Sarpi B, Rioult M, Aghavnian T, Moussy JB *et al.* Electrostriction, Electroresistance, and Electromigration in Epitaxial $BaTiO_3$-Based Heterostructures: Role of Interfaces and Electric Poling. ACS Appl. Nano Mater. 2019; **2**, 3556-3569.

[41] Khan S, Humera N, Niaz S, Riaz S, Atiq S, Naseem S. Simultaneous Normal – Anomalous Dielectric Dispersion and Room Temperature Ferroelectricity in CBD Perovskite $BaTiO_3$ Thin Films. J Mater Res Technol. 2020; **9**, 11439-11452.

[42] Dai YQ, Dai JM, Tang XW, Zhang KJ, Zhu XB, Yang J *et al.* Thickness Effect on the Properties of $BaTiO_3$-$CoFe_2O_4$ Multilayer Thin Films Prepared by Chemical Solution Deposition. J. Alloys Compd. 2014; **587**, 681-687.

[43] Alberca A, Munuera C, Azpeitia J, Kirby B, Nemes NM, Perez-Munoz AM, Tornos J, Mompean FJ, Leon C, Santamaria J, Garcia-Hernandez M. Phase Separation Enhanced Magneto-Electric Coupling in $La_{0.7}Ca_{0.3}MnO_3/BiTiO_3$ Ultra-Thin Films. Sci Rep. 2015; **5**, 17926.

[44] Jiang G, Zhou D, Yang J, Shao H, Xue F, Fu Q. Microstructure and Multiferroic Properties of $BaTiO_3/CoFe_2O_4$ Films on Al_2O_3/Pt Substrates Fabricated by Electrophoretic Deposition. J. Eur. Cer. Soc. 2013; **33**, 1155-1163.

[45] Erdem D, Bingham NS, Heiligtag FJ, Pilet N, Warnicke P, Vaz CAF *et al.* Nanoparticle-Based Magnetoelectric $BaTiO_3$– $CoFe_2O_4$ Thin Film Heterostructures for Voltage Control of Magnetism. ACS Nano. 2016; **10**, 9840-9851.

[46] Tang Y, Wang R, Zhang Y, Xiao B, Li S, Du P. Magnetoelectric Coupling Tailored by the Orientation of the Nanocrystals in One Component in Percolative Multiferroic Composites. RSC Adv. 2019; **9**, 20345.

[47] Mudiyanselage RRHH, Magill BA, Burton J, Miller M, Spencer J, McMillan

K *et al.* Coherent Acoustic Phonons and Ultrafast Carrier Dynamics in Hetero-Epitaxial $BaTiO_3$–$BiFeO_3$ Films and Nanorods. J. Mater. Chem. C. 2019; 7, 14212.

[48] Yan LH, Liang WZ, Liu SH, W Huang, Lin Y. Multiferroic $BaTiO_3$-$CoFe_2O_4$ Nano Composite Thin Films Grown by Polymer-Assisted Deposition. Integ. Ferr. 2011; **131**, 82-88.

[49] Gao X, Zhang D, Wang X, Jian J, He Z, Dou H, Wang H. Vertically Aligned Nanocomposite $(BaTiO_3)_{0.8}$: $(La_{0.7}Sr_{0.3}MnO_3)_{0.2}$ Thin Films with Anisotropic Multifunctionalities. Nanoscale Adv. 2020; **2**, 3276.

[50] Sun Y, Zhang L, Wang H, Guo M, Lou X, Wang D. Excellent Thermal Stability of Large Polarization in $(Bi_{0.5}Na_{0.5})$ TiO_3-$BaTiO_3$ Thin Films Induced by Defect Dipole. Appl. Surf. Sci. 2019; **504**, 144391.

[51] Kawamura G, Oura K, Tan WK, Goto T, Nakamura Y, Yokoe D *et al.* Nanotube Array-Based Barium Titanate-Cobalt Ferrite Composite Film for Affordable Magnetoelectric Multiferroics. J. Mater. Chem. C. 2019; 7, 10066-10072.

[52] Zhou L, Fu Q, Xue F, Tang X, Zhou D, Tian Y *et al.* Multiple Interfacial Fe_3O_4@$BaTiO_3$/P(VDF-HFP) Core-Shell-Matrix Films with Internal Barrier Layer Capacitor (IBLC) Effects and High Energy Storage Density. ACS Appl. Mater. Interf. 2017; **9**, 40792-40800.

[53] Verma KC, Kotnala RK, Ram M, Negi NS. Optical Characteristics of Nanocrystalline $Pb_{1-x}Sr_xTiO_3$ Thin Films Using Transmission Measurements. Adv. Sci. Focus 2012; 1(1), 1-7.

[54] Verma KC, Shah J, Kotnala RK. Strong Magnetoelectric Coupling of $Pb_{1-x}Sr_x(Fe_{0.012}Ti_{0.988})O_3$ Nanoparticles. J. Nanosci. Nanotechnol. 2015; 15(2), 1587-1590.

[55] Bhatti HS, Hussain ST, Khan FA, Hussain S. Synthesis and Induced Multiferroicity of Perovskite $PbTiO_3$; A review. Appl. Surf. Sci. 2016; **367**, 291-306.

[56] Verma KC, Negi NS. Magnetoelectric coupling and relaxor ferroelectric properties of $Pb_{0.7}Sr_{0.3}(Fe_{0.012}Ti_{0.988})O_3$ thin film. Scr. Mater. 2010; **63**, 891-894.

[57] Verma KC, Singh M, Kotnala RK, Negi NS. Ferromagnetism and ferroelectricity in highly resistive $Pb_{0.7}Sr_{0.3}(Fe_{0.012}Ti_{0.988})O_3$ nanoparticles and its conduction by variable-range-hopping mechanism. Appl. Phys. Lett. 2008; 93, 072904.

[58] Sakamoto W, Iwata A, Moriya M, Yogo T. Electrical and magnetic properties of Mn-Doped $0.7BiFeO_3$-$0.3PbTiO_3$ Thin Films Prepared Under Various Heating Atmospheres. Mater. Chem. Phys. 2009; **116**, 536-541.

[59] Gopman DB, Chen P, Lau JW, Chavez AC, Carman GP, Finkel P *et al.* Large Interfacial Magnetostriction in $(Co/Ni)_4$/ $Pb(Mg_{1/3}Nb_{2/3})O_3$–$PbTiO_3$ Multiferroic Heterostructures. ACS Appl. Mater. Interf. 2018; **10**, 24725-24732.

[60] Zhang P, Gao C, Lv F, Wei Y, Dong C, Jia C *et al.* Hydrothermal Epitaxial Growth and Nonvolatile Bipolar Resistive Switching Behavior of $LaFeO_3$-$PbTiO_3$ Films on Nb: $SrTiO_3(001)$ Substrate. Appl. Phys. Lett. 2014; **105**, 152904.

[61] Murakami M, Chang KS, Aronova MA, Lin CL, Yu MH, Simpers JH *et al.* Tunable Multiferroic Properties in Nanocomposite $PbTiO_3$-$CoFe_2O_4$ Epitaxial Thin Films. Appl. Phys. Lett. 2005; **87**, 112901.

[62] Pan M, Liu Y, Bai G, Hong S, Dravid VP, Petford-Long AK. Structure-Property Relationships in Self-

Assembled Metalorganic Chemical Vapor Deposition–Grown $CoFe_2O_4$–$PbTiO_3$ Multiferroic Nanocomposites Using Three Dimensional Characterization. Appl. Phys. Lett. 2011; **110**, 034103.

[63] Feng M, Hu J, Wang J, Li Z, Shu L, Nan CW. Converse Magnetoelectric Coupling in Ni Fe/Pb $(Mg_{1/3}Nb_{2/3})$ O_3–$PbTiO_3$ Nanocomposite Thin Films Grown on Si Substrates. Appl. Phys. Lett. 2013; **103**, 192903.

[64] Levin I, Li J, Slutsker J, Roytburd AL. Design of Self-Assembled Multiferroic Nanostructures in Epitaxial Films. Adv. Mater. 2006; **18**, 2044-2047.

[65] Zhu L, Dong Y, Zhang X, Yao Y, Weng W, Han G et al. Microstructure and Properties of Sol-Gel Derived $PbTiO_3/NiFe_2O_4$ Multiferroic Composite Thin Film with the Two Nano-Crystalline Phases Dispersed Homogeneously. J. Alloys Compd. 2010; **503**, 426-430.

[66] Kotnala RK, Verma KC, Mathpal MC, Negi NS. Multifunctional behaviour of nanostructured $Pb_{0.7}Sr_{0.3}(Fe_{0.012}Ti_{0.988})O_3$ thin film. J. Phys. D: Appl. Phys. 2009; **42**, 085408.

[67] Haislmaier RC, Lu Y, Lapano J, Zhou H, Alem N, Sinnott SB *et al.* Large tetragonality and room temperature ferroelectricity in compressively strained $CaTiO_3$ thin films. APL Mater. 2019; 7, 051104.

[68] Karpinsky DV, Troyanchuk IO, Sikolenko V, Efimov V, Kholkin AL. Electromechanical and magnetic properties of $BiFeO_3$-$LaFeO_3$-$CaTiO_3$ ceramics near the rhombohedral-orthorhombic phase boundary. J. Appl. Phys. 2013; **113**, 187218.

[69] Jia T, Fan Z, Yao J, Liu C, Li Y, Yu J *et al.* Multi-Field Control of Domains in A Room Temperature Multiferroic $0.85BiTi_{0.1}Fe_{0.8}Mg_{0.1}O_3$-$0.15CaTiO_3$ Thin Film. ACS Appl. Mater. Interf. 2018; 10(**24**), 20712-20719.

[70] Wu X, Kuang D, Yao L, Yang S, Zhang Y. The Structural, Optical, Ferroelectric Properties of (1-x) $BiFeO_3$-$xCaTiO_3$ Thin Films by a Sol–Gel Method. J Mater Sci: Mater Electron. 2017; 28, 493-500.

[71] Liu C, An F, Gharavi PSM, Lu Q, Zha J, Chen C *et al.* Large-Scale Multiferroic Complex Oxide Epitaxy with Magnetically Switched Polarization Enabled by Solution Processing, Natl. Sci. Rev. 2020; 7, 84-91.

[72] LV F, Gao C, Zhang P, Dong C, Zhang C, Xue D. Bipolar Resistive Switching Behavior of $CaTiO_3$ Films Grown by Hydrothermal Epitaxy. RSC Adv. 2015; **5**, 40714-40718.

Chapter 4

Research Progress of Ionic Thermoelectric Materials for Energy Harvesting

Jianwei Zhang, Ying Xiao, Bowei Lei, Gengyuan Liang and Wenshu Zhao

Abstract

Thermoelectric material is a kind of functional material that can mutually convert heat energy and electric energy. It can convert low-grade heat energy (less than 130°C) into electric energy. Compared with traditional electronic thermoelectric materials, ionic thermoelectric materials have higher performance. The Seebeck coefficient can generate 2–3 orders of magnitude higher ionic thermoelectric potential than electronic thermoelectric materials, so it has good application prospects in small thermoelectric generators and solar power generation. According to the thermoelectric conversion mechanism, ionic thermoelectric materials can be divided into ionic thermoelectric materials based on the Soret effect and thermocouple effect. They are widely used in pyrogen batteries and ionic thermoelectric capacitors. The latest two types of ionic thermoelectric materials are in this article. The research progress is explained, and the problems and challenges of ionic thermoelectric materials and the future development direction are also put forward.

Keywords: thermoelectric materials, pyrogen batteries, ionic thermoelectric capacitors

1. Introduction

Thermoelectric material is a kind of functional material that can mutually convert heat energy and electric energy. It can convert low-grade heat energy below 130°C into electric energy, so it has good application prospects in small thermoelectric generators and solar power generation. Traditional electronic thermoelectric materials are divided into inorganic thermoelectric materials and organic thermoelectric materials: inorganic thermoelectric materials include Bi2Te3 [1], PbTe [2], SiGe [3], etc. and their alloys, due to their high thermal conductivity, mechanical strength, and thermoelectric conversion efficiency, they have been widely used in thermoelectric coolers, thermoelectric generators, etc., but there are problems such as high price, complicated processing technology and easy to produce heavy metal pollution; organic thermoelectric materials is a kind of conductive polymer with conjugate structure polymers, mainly polyaniline, polypyrrole, polythiophene and its derivatives [4], etc., this kind of conductive high molecular polymer has good flexibility, low price, safety and pollution-free, and gradually replaces inorganic thermoelectric materials into people's field of vision. However, the Seebeck coefficient of organic thermoelectric materials is very low, there are problems such as low conductivity and power factor, low thermoelectric conversion efficiency, etc., usually need doping modification, composite with carbon materials [5], inorganic thermoelectric materials [6] and other methods to improve their thermoelectric performance.

With the development of materials science, a new type of thermoelectric material-ionic thermoelectric material has gradually entered people's field of vision. Compared with traditional electronic thermoelectric materials, ionic thermoelectric materials have a higher Seebeck coefficient (usually several hundred times higher than traditional thermoelectric materials) and can generate 2–3 orders of magnitude higher ion thermoelectric potential than electronic thermoelectric materials, such materials have a wide range of applications in pyrogen batteries and ion thermoelectric capacitors.

The evaluation of the thermoelectric performance of thermoelectric materials is usually expressed by the dimensionless thermoelectric figure of merit ZT:

$$ZT = \frac{\sigma S^2}{\kappa} \tag{1}$$

Among them, T is the absolute temperature, S is the Seebeck coefficient, which represents the inherent electron transport performance of thermoelectric materials, and is also the most basic parameter that determines the thermoelectric effect of the material; σ is the electrical conductivity, which represents the current transport capacity of the thermoelectric material; κ is the thermal conductivity, which is a performance parameter to measure the heat transfer of materials. Thermoelectric materials with excellent performance should have high electrical conductivity and Seebeck coefficient, as well as low thermal conductivity. In addition, when investigating the conversion efficiency of thermoelectric generators, the power factor $PF = \sigma S^2$ will also be used.

The thermoelectric power of traditional electronic thermoelectric conversion materials comes from the migration of internal electrons under a temperature gradient, while the thermoelectric power of ionic thermoelectric materials comes from ion migration under temperature differences. This effect is also called the ion Seebeck effect, or Soret effect, the made thermoelectric device is called an ion thermoelectric capacitor. In addition to this way of generating thermoelectric potential, ionic thermoelectric materials have another source of thermoelectric potential: when a redox pair is used in the material, redox reactions occur on both sides of the material electrode, and the entropy change in the reaction is used to achieve heat energy to electrical energy. This effect is called the thermocouple effect, and the thermoelectric conversion device made is called pyrogen cell.

This article reviews the research progress of ionic thermoelectric materials based on the Soret effect and thermocouple effect, combs the thermoelectric conversion mechanism of such ionic thermoelectric materials used in energy conversion systems, and summarizes the research of various new electrolytes and electrode materials. Progress. Finally, the future development direction of ionic thermoelectric materials is prospected.

2. Ionic thermoelectric materials for pyrogen batteries

Ionic thermoelectric materials with added redox couples mainly use the temperature effect between the redox pairs, which is the thermocouple effect.

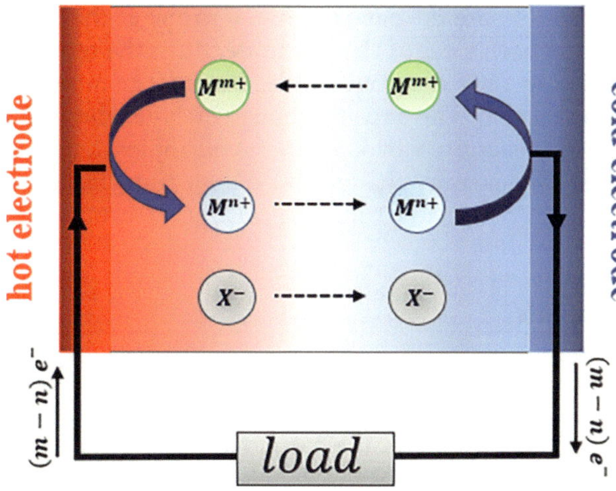

Figure 1.
Working mechanism of TGC device.

The redox reaction occurs between the redox couple and the electrode in the material electrolyte, and the temperature gradient promotes the two types of electrons. The migration and diffusion of the opposite electrodes use the entropy change of the reaction to realize the conversion from thermal energy to electrical energy. The thermoelectric conversion device made based on this redox pair is called a thermogenic cell (TGC).

2.1 Thermogenic battery (TGC)

The redox-active electrolyte in the pyrogen cell undergoes oxidation/reduction reactions on the electrodes on both sides to produce aggregation, and spontaneously moves to the other electrode under the concentration and temperature gradient, while the electrons generated by the reaction can enter and exit the electrodes to achieve current flow continuous output, the pyrogen battery has the advantages of low maintenance cost and long-term stable operation.

The thermogenic battery consists of two identical electrodes and an electrolyte containing a redox pair, and its operating mechanism is shown in the **Figure 1** below. When the temperature of the electrodes on both sides is inconsistent, the redox pair (such as M^{m+}, M^{n+}) moves to both sides of the electrode under the action of the temperature gradient, and oxidation/reduction reactions occur respectively at the electrode/electrolyte interface, and the electrons lost or gained in the reaction. The directional migration occurs through the external circuit to form a current. At the same time, the ions gathered around the electrode after the reaction will migrate to the other electrode under the action of the concentration gradient through the electrolyte solution and continue to undergo the reduction/oxidation reaction, thereby maintaining the continuity of the current output in the entire battery, and the two ends of the electrode are maintained stable and constant potential difference, the magnitude of the potential difference depends on the entropy value of the redox pair.

In the thermogenic battery, the Seebeck coefficient is determined by the partial molar entropy difference of the redox pair [7]:

$$S_e = \frac{\Delta V}{\Delta T} = \frac{S_B - S_A}{nF} \tag{2}$$

Among them, S_A and S_B are the partial molar entropy of the redox pair, ΔV is the open-circuit voltage on both sides of the electrode, ΔT is the temperature difference between the hot and cold ends of the electrode, and n is the number of electrons transferred in a unit reaction. The entropy change includes the entropy change caused by the electrode reaction, various transfer entropies, and the transmission entropy of electrons. This entropy change will change significantly with the properties of the redox couple. At the same time, the solvent environment also affects the entropy change.

2.2 Electrolyte materials

At present, the most widely used electrolyte material in thermogenic batteries is gel electrolyte, which has high ionic conductivity and flexibility and can also hinder the overall heat conduction of the battery and prevent electrolyte leakage. Gel electrolyte is usually composed of polymer matrix and electrolyte added with redox pair. The cross-linking of the polymer provides mechanical strength and a certain network support space, in which redox ions migrate freely and move to the electrodes on both sides for redox reaction. Ionic thermoelectric materials are classified into p-type and n-type thermoelectric materials according to the positive and negative of the ion Seebeck coefficient, and the redox pairs contained therein are also called p-type and n-type redox pairs, respectively.

In order to effectively collect waste heat, stacking individual thermocouples together to form a battery-like series structure can increase the potential output of the thermogenic battery exponentially. However, if only a single type of redox couple is used, a hot-cold-hot short circuit between adjacent units will occur (as shown in **Figure 2**). Take the p-type redox couple $[Fe(CN)_6]^{3-}$ /$[Fe(CN)_6]^{4-}$ as an example, If the same type of thermocouple is used, the hot end of the thermocouple must be connected to the cold end of the next thermocouple by wire, which will cause unnecessary heat transfer between the hot end and the cold end. Therefore, the TEC device is made by connecting n-type thermocouples and p-type thermocouples in series, and the polarity of the battery can be replaced by using alternating n-type thermocouples and p-type thermocouples, so as to reduce unnecessary heat transfer, simplify the manufacturing process of the device and realize the effective utilization of heat energy.

2.2.1 n-type redox couple

Commonly used n-type redox pairs include $Co(bpy)_3^{2+/3+}$, $Fe^{2+/3+}$, I^-/I^{3-}, etc., and their Seebeck coefficients are negative. $Co(bpy)_3^{2+/3+}$ is a more commonly used

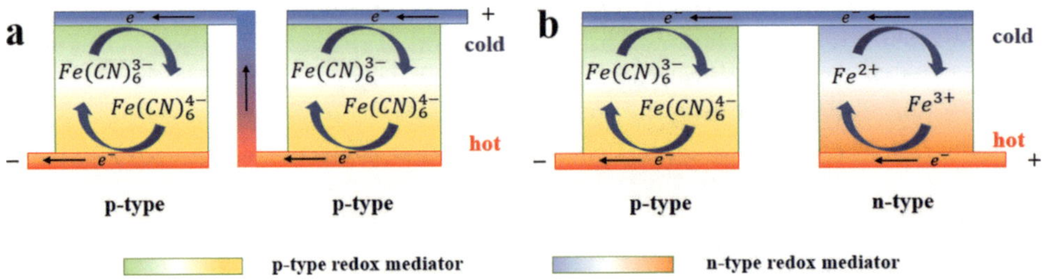

Figure 2.
Hot-cold-hot short-circuit phenomenon of TGC. a) Illustration of two in-series-connected p-type half-cells that both use aqueous $Fe(CN)_6^{4-}/Fe(CN)_6^{3-}$ as the redox mediator. b) Illustration of a p-type half-cell (using aqueous $Fe(CN)_6^{4-}/Fe(CN)_6^{3-}$ as the redox mediator) and an n-type half-cell (using aqueous Fe^{2+}/Fe^{3+} as the redox mediator) that are connected in series.

n-type redox pair. In the redox reaction, electron transfer will cause changes in the spin state of the central ion $Co^{2+/3+}$, causing additional entropy effects. This redox pair has low solubility in water and is often used in organic solvents or ionic liquid electrolytes. $Co^{II/III}(bpy)_3(NTf_2)_{2/3}$ has a higher Seebeck coefficient in organic solvents. Jiao et al. first tried to use $Co^{II/III}(bpy)_3(NTf_2)_{2/3}$ as a redox couple, and It has been applied to five different ionic liquid systems [8], and found that it has lower thermal conductivity in the $[C_{2mim}][NTf_2]$ system, which is 0.16 W/(m·K). When using 0.1 M $Co^{II/III}$ $(bpy)_3(NTf_2)_{2/3}$ pairs, the maximum output power of TEC reaches 77 mW/m^2.

In addition to ionic liquids, gelling agents can also be used to prepare quasi-solid gel electrolytes to prevent electrolyte leakage. Pringle et al. used polyvinylidene fluoride (PVDF) and polyvinylidene fluoride co-fluoropentene (PVDF-HFP) as gelling agents [9], using a hybrid gel method to prepare gel electrolytes. By optimizing the concentration of redox pairs and the interval between cold and hot electrodes, the performance of the thermoelectric battery was further improved. On this basis, Pringle et al. added 5% biphasic cellulose and used a solvent exchange method to prepare an electrolyte gel containing $Co^{II/III}(bpy)_3(NTf_2)_{2/3}$ redox couple [10]. The use of this cellulose structure as the transmission network of redox couples avoids the toxicity of organic solvents and the risk of flammability and leakage. Cellulose also provides mechanical flexibility for gel electrolytes and provides new ideas for the development of human wearable devices. In addition, Laux et al. combined the $Co^{2+/3+}$ $(bpy)_3$ $(TFSI_2)_{2/3}$ redox pair with choline lactate, and its Seebeck coefficient was increased to −3.63 mV/K [11].

Another type of n-type redox couple is $Fe^{2+/3+}$, which has good chemical stability in water and is also used as a redox couple for thermogenic battery electrolytes. Polyvinyl alcohol (PVA) has good mechanical tensile properties, non-toxic and environmentally friendly, hydrophilic and high-temperature resistant, and has low production cost. Fe^{3+} makes the solution strongly acidic under the action of hydrolysis, which can just maintain the acidic gel environment required by PVA, so PVA is often added to $Fe^{2+/3+}$ to prepare gel electrolyte. Zhou et al. prepared a PVA-$FeCl_{2/3}$ film [12], the power density of which exceeds the Co redox couple. On this basis, he also added HCl solution [13], which significantly improved the mechanical tensile properties of the gel and synergistically reduced the charge transfer resistance. This type of gel has good thermoelectrochemical properties. In addition, phosphoric acid (H_3PO_4) can also significantly increase the ionic conductivity of PVA-based polymer electrolytes. This type of gel electrolyte has broad application prospects in supercapacitors, fuel cells, and lithium batteries.

2.2.2 p-type redox pair

Zhou et al. used PVA as a gel solution and added $FeCl_3/FeCl_2$ or potassium ferricyanide/potassium ferrocyanide as two redox pairs to obtain a quasi-solid ion gel with positive (PFC) or negative (PPF) Seebeck coefficients. Glue [12], and fabricated an integrated wearable thermal battery, this thermal battery has excellent mechanical properties and thermoelectric properties can be used for low-level thermal energy conversion. The mechanism of PFC and PPF gel generating thermoelectric voltage is similar to the transfer mechanism of pure Fe^{2+}/Fe^{3+} and $Fe(CN)_6{}^{4-}/Fe(CN)_6{}^{3-}$ in an aqueous solution. The charge is not through the migration of electrons or holes in the conductor, but it is the migration of ions under the action of temperature difference. This integrated wearable device can use body heat to generate an open circuit voltage of about 0.7 V and a short circuit current of 2 mA and achieve a maximum output power of about 0.3 mW.

The above-mentioned $Fe(CN)_6^{4-}/Fe(CN)_6^{3-}$ is the most typical p-type redox pair. Aldous et al. studied the water-based thermal battery containing $K_3Fe(CN)_6/K_4Fe(CN)_6$ as gel electrolyte and compared the interaction between traditional gels such as SiO_2 nanoparticles and agar, and new gelling materials of sodium polyacrylate beads, and the aqueous solution system [14]. This kind of three-dimensional network crosslinked polymer is self-assembled by hydrogen bonding and hydrophobic interaction, which can fix electrolyte solution in situ. Experiments show that fumed silica cannot form an effective gel in a neutral solution and is mainly used in a strongly acidic gel electrolyte. The effective, uniform, and transparent gel electrolyte can be formed by mixing 5.5 wt% sodium polyacrylate beads in 0.1 M $Fe(CN)_6^{3-/4-}$ solution and the maximum efficiency of thermocouple is twice higher than that of the liquid thermocouple.

Cellulose is an environmentally friendly biomaterial, and it can also be used as the curing matrix of liquid electrolytes. Cellulose is insoluble in water, and the porous structure of its fiber network provides additional heat transfer paths and tortuous ion diffusion channels. Pringle et al. soaked cellulose membrane in 0.4 M $Fe(CN)_6^{3-/4-}$ redox electrolyte, and measured that the power density of TEC can reach 168 mW/m^2, and increasing the concentration of ionic liquid can further improve the current density [10]. Using this kind of cellulose-based electrolyte can improve the safety of equipment and significantly reduce the amount of electrolyte.

Gelatin can also be used as an electrolyte matrix for fixing redox couple because of its low cost, good biocompatibility, and good mechanical flexibility. Han et al. prepared Gelatin-KCl-$Fe(CN)_6^{3-/4-}$ ionic gel electrolyte in gelatin matrix with KCl, NaCl, and KNO_3 as ion donors [15], and showed a high thermal potential of 17.0 mv/k. 25 components were made into wearable equipment, which could generate a high output voltage of 2.2 V by using the temperature difference between the human body and environment. This kind of ionic thermoelectric material has a synergistic effect, and the ion provider realizes the thermal diffusion effect. KCl accumulates positive charges near the cold end of the electrode through thermal diffusion and generates an electric field from the cold end to the hot end, which is called thermal diffusion voltage. The reduction pair realizes the thermocurrent effect. $FeCN^{3-}$ is more likely to undergo oxidation reaction than $FeCN^{4-}$ under high-temperature electrode, and produces higher solvation entropy, and electrons migrate to the hot end electrode, thus improving the electrochemical potential. $FeCN^{3-}$ gets electrons at the cold end to generate $FeCN^{4-}$, which drives thermal diffusion and balances oxidation–reduction reaction under the temperature gradient. The thermoelectric current effect makes the potential of electrodes at both ends move in the same direction as the thermal diffusion effect. $FeCN^{4-/3-}$ also participates in thermal diffusion and contributes to the final thermoelectric energy (**Figure 3**).

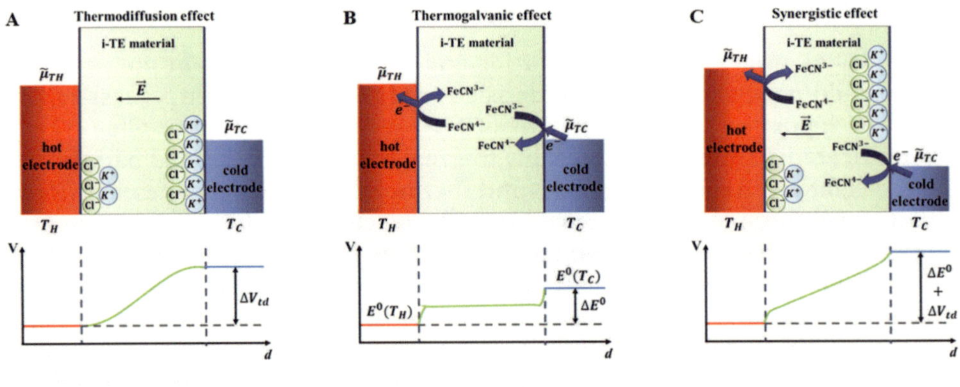

Figure 3.
Mechanism of the synergistic effect. Electrochemical potential (μ~) of charge carries diagrams and the corresponding voltage (V) distribution of i-TE material of Gelatin-KCl-FeCN.

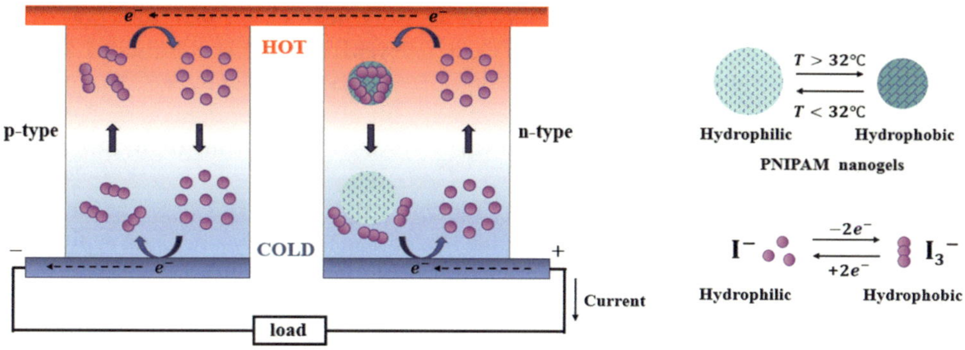

Figure 4.
Schematic of the p–n conversion mechanism for the I^-/I^-_3 system (left). The hydrophilic/hydrophobic conversion of PNIPAM nanogels and the equilibrium3 reaction between I^- and I^-_3 (right).

2.2.3 p-n conversion of redox pair

Using the special properties of some solvents, the reversion of redox to N-type and P-type can also be realized. I^-/I_3^- is a redox couple with a positive Seebeck coefficient, but Zhou et al. induced the redox couple of I^-/I_3^- to undergo p-n conversion by using the heat-sensitive polymer of n-isopropylacrylamide (PNIPAM), and prepared gel electrolyte with negative Seebeck coefficient [16]. The mechanism of p-n conversion is that the transition temperature of PNIPAM nanogels synthesized by emulsion polymerization is about 32°C, and above this temperature, the gel changes from hydrophilic state to hydrophobic state. In these PNIPAM nanogels, relatively hydrophobic I can be captured by dehydrated nanogels at the hot end by hydrophobic action, and electrons are released at the hot end, resulting in free I_3 passing through the battery with a concentration gradient, resulting in Seebeck coefficient changing from 0.71 to −1.91 mV/K. The transition temperature of 32°C is very suitable for human body temperature, which is especially suitable for human body heat collection. Zhou et al. prepared a wearable device composed of alternating I^-/I_3^- (p-type) and PNIPAM- I^-/I_3^- gel (n-type) units connected in series. The open-circuit voltage of about 1 V was generated by human body heat energy, and the maximum output power reached 9 μW, which can meet the power consumption requirements of small wearable devices. This work opens a new perspective for controlling the Seebeck coefficient of redox pairs in a gel electrolyte (**Figure 4**).

2.3 Electrode materials

Metal electrode material is the earliest electrode type used in the thermal cell, and the platinum electrode is a typical non-reactive catalytic material, which is widely used because it does not participate in electrode reaction and has high conductivity. However, platinum, as a precious metal, has a high production cost, so developing electrode materials with low price and higher conductivity is a thinking direction of thermoelectric chemical batteries at present.

The porous structure of nano-carbon materials makes the electrode have a higher active surface area and can generate higher thermal voltage under thermal gradient. Among them, carbon nanotubes (CNTs) and reduced graphene oxide (r-GO) are ideal candidate materials for thermoelectric battery electrodes. CNTs have fast carrier migration, excellent chemical stability, and catalytic activity. Kim et al. deposited CNTs on activated carbon textiles by dip-coating method [17] to obtain higher specific surface area and flexibility. The thermocouple manufactured can generate the maximum power density of 0.46 mW/m^2 by bulk heat. Chen et al. vacuum filtered the r-GO dispersion to obtain an electrode with a thickness of about 1.8 μm, which was applied in an electrolyte solution of 0.4 M $K_3Fe(CN)_6$/ $K_4Fe(CN)_6$. when the temperature difference was 60°C, the maximum mass power density of the r-GO electrode reached 25.51 W/kg, which was three orders of magnitude higher than that of the platinum electrode [18].

Pure CNTs have tortuous transmission channels, while pure r-GO is easy to form sheet-like accumulation, which is unfavorable to electron transmission. Chen et al. prepared SWNT-rGO composite electrode material by combining SWNT with reduced graphene oxide dispersed in 1- cyclohexyl –2- pyrrolidone (CHP) [19], and designed porosity for thermocouple electrode by controlling electrode composition and thickness. Increasing the amount of reduced graphene oxide will lead to the sheet stacking again, while decreasing the amount of r-GO will also lead to a more tortuous structure and hinder ion movement, thus reducing the performance of thermocouples. By optimizing the composition and thickness of the electrode, the balance between porosity and curvature is obtained, which is beneficial to the rapid diffusion of electrolytes. When 90% SWNT and 10% r-GO are added, the electrode thermocouple has the best performance, and when the temperature difference is t = 31°C, the power density is 0.327 W/m^2.

Apart from pure carbon-based materials, the conductive polymer is another promising flexible electrode active material, which not only has high polymer porosity and high active surface area but also shows good flexibility and stretchability. Poly (3,4-ethylenedioxythiophene)/polystyrene sulfonic acid (PEDOT: PSS) is a promising conductive polymer. When it is added to graphene or carbon nanotube films, it can alleviate the bending of carbon nanotubes and mass transfer problems caused by graphene sheet stacking. Chen et al. prepared a composite thin film electrode composed of PEDOT: PSS, edge functionalized graphene (EFG), and carbon nanotubes by a simple drop-casting method [20]. PEDOT: PSS can not only provide an ion transport pathway and increase the porosity of EFG/ CNT, but also be used as a dispersant and film-forming agent in the manufacturing process to promote the uniform dispersion of EFG/CNT in the electrode, which in turn provides a more catalytic surface in the composite membrane.

3. Ionic thermoelectric materials for ionic thermoelectric capacitors

Due to the temperature gradient, the anions and cations in the thermoelectric electrolyte migrate directionally to both sides of the electrode to form a potential difference, which is then connected with an external circuit to generate a thermal voltage, thus realizing the conversion from thermal energy to electrical energy. This thermoelectric conversion device based on the Soret effect is called an ion thermoelectric capacitor (TEC).

3.1 Ion thermoelectric capacitor (TEC)

Ion thermoelectric capacitor is composed of two identical electrodes and a solid electrolyte containing positive and negative ions. Driven by the temperature difference between the two ends, the directional migration of internal ions is realized due to the Soret effect. When the ions gather around the electrodes, the continuous generation of current in the circuit is realized because the ions cannot enter the electrodes. Similar to a capacitor, after the temperature difference is generated, with the continuous accumulation of ions at both ends of the electrodes, the current will gradually decrease to reach equilibrium, and the maximum thermal potential will be generated at both ends of the electrodes at this time. This working mode is discontinuous, and the current

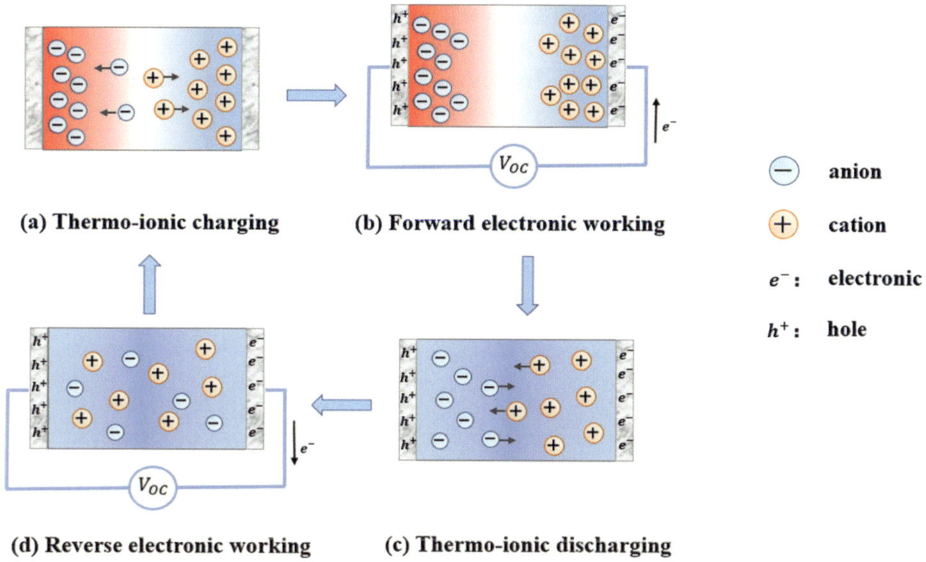

Figure 5.
Working mechanism of TEC device.

output mode belongs to transient output, which requires human intervention to realize the storage of thermal energy into electrical energy. This device is mainly used for intermittent heat sources, such as solar cells and supercapacitors. The working mechanism of TEC is shown in the **Figure 5** [21], which is mainly divided into four stages.

1. Thermionic charging stage: One end of the electrode is heated, resulting in a temperature difference between the two electrodes. The anion and cation migrate to both ends of the electrode due to the Soret effect, and the final gathering produces a voltage difference between the two electrodes, which is called ion thermoelectric potential.

2. Forward electronic working stage: When the electrodes on both sides are connected with the external load, the electrons in the circuit migrate through the external load, and finally, the holes and electrons are collected on the electrodes on both sides to achieve charge balance. The current generated at this stage can do work for the external load.

3. Thermionic discharge stage: When the external load is disconnected and the heating end of the ion thermoelectric capacitor is cooled, the disappearance of the temperature gradient will drive the ions back to their original state, but the electrons remain on the electrodes on both sides, and there is still a voltage difference between the two ends of the electrodes.

4. Reverse electronic working stage: When the external load is connected again, electrons flow out of the capacitor electrode through the external circuit, thus doing work for the external load.

In ion thermoelectric capacitors, the Seebeck coefficient is related to the Soret effect. Different types of particles have different thermal diffusion in the mixture, and the electrode produced when particles migrate from the hot end to the cold end is called the positive electrode. On the contrary, the electrode produced when particles migrate from the cold end to the hot end is called the negative electrode.

Under the influence of the Soret effect, positive and negative ions diffuse to the electrodes on both sides in opposite directions, and then a voltage gradient, that is, thermal voltage, will be generated. Its contribution to Se is expressed as [22]:

$$S_e = \frac{F}{\sigma T} zcD_T \tag{3}$$

In the formula, z is the number of ion charges, F is the Faraday constant, c is the ion concentration, mol/m^3; σ is the conductivity, S/m; D_T is the Soret thermal diffusion coefficient of the ion.

3.2 Electrolyte material

The electrolyte in the ionic thermoelectric capacitor is mainly composed of an organic polymer and a salt electrolyte that can be electrolyzed into ions and has ionic conductivity. The polymer molecular chains are entangled with each other to form a certain network space structure. Anions and cations are connected with the functional groups of the polymer by coordination bonds and are distributed in the gaps of the polymer structure. When there is a temperature gradient, the anions and cations Due to the directional migration of the Soret effect, an uneven charge distribution is generated inside and a thermal voltage is generated. Common electrolytes can be divided into conductive polymers, single-ion conductors, ionic liquid quasi-solid gels, and chemically cross-linked ionic gels according to their structure and composition.

3.2.1 Conductive polymer

The most widely used electrolyte materials for thermionic capacitors are conductive polymer electrolytes. Among them, PEDOT: PSS has become the most in-depth researched type of conductive polymer material due to its excellent conductivity, good solution processing performance, and low price. Crispin researched PEDOT: PSS for the first time, and synthesized various PEDOT derivatives (PEDOT: PSS, PEDOT: PSS: PSSNA, etc.), which provided the basis for the electrolyte materials of thermionic capacitors [23]. PEDOT: PSS is a kind of mixed ion electronic conductor. Both electrons and ions can be used as charge carriers in electrolyte materials. It is the basic material of many emerging ionic thermoelectric materials. This type of conductive polymer has two types of thermoelectric current and thermovoltage. Contribution: The constant contribution of electrons and the non-constant contribution of ions. The total thermoelectric energy is expressed as the sum of the electron Seebeck coefficient and the ion Seebeck coefficient. For PEDOT: PSS polymer mixed conductors, the ion Seebeck effect accounts for the main part. The maximum power factor is increased by 2 to 4 orders of magnitude. Therefore, Crispin added PSSNa to PEDOT: PSS to enhance ionic conductivity and increase the mobility of cations. Relative humidity also has a significant effect on ion transmission. The results show that the Seebeck coefficient suddenly rises when the relative humidity is about 40%, and finally reaches 215 μV/K at a relative humidity of 80%.

In addition, Kim et al. used PSSH as an additive, doped it into PEDOT: PSS polymer to make an ion hybrid thermoelectric capacitor [24], and realized ion thermally driven diffusion through polystyrene sulfonic acid (PSSH) film, which greatly improved the output voltage (8 mV/K). If a polyaniline coating containing graphene and carbon nanotubes is used as an electrode, a large area capacitance

(1200 F/m^2) of 38 mV can be generated at a small temperature difference (5 K). The exploration of this thermal drive ion diffusion behavior provides a new research direction for the collection of low-grade heat energy.

3.2.2 Single ion conductor

In addition to the self-contained cations in the conductive polymer, it is also possible to add an electrolyte solution containing free cations to the PEDOT:PSS to prepare a single-ion conductor thermoelectric material, which can further improve the conductivity of the electrolyte material, thereby increasing the thermoelectricity of the ion thermoelectric capacitor Figure of merit and power factor, in which Ag ions and Na ions are the most used ion transport carriers.

Segalman et al. prepared Ag:PSS by ion exchange, and then incorporated it into PEDOT:PSS to form a conductive mixture enriched with Ag:PSS along PEDOT [22], in which PEDOT:PSS was used as an electronic conductor and Ag^+ was used as a carrier carrying out ion transport, but also participating in the electrochemical reaction of the silver electrode. This mixed ion-electronic conductive polymer combines the redox reaction on the electrode and the thermoelectric transport of ions in the system. It is an electrochemically active material that enhances the Seebeck coefficient of ions. However, this material requires anions to pass from an electrode in one direction. Transmitted to the other electrode, if all the silver at one end of the system is consumed, the ion Seebeck enhancement will disappear.

Nafion is an ion exchanger composed of perfluorinated sulfonate. There are hydrophobic fluorocarbon framework domains and hydrophilic domains dominated by ionized sulfonic acid groups in its molecules. It has good cation selectivity and excellent ionic conductivity; it is often used as a cation exchange membrane. At the same time, it also has strong resistance to chemical corrosion. The working temperature is higher, which can reach about 190°C. Therefore, this perfluorosulfonic acid polymer is widely regarded as the best polymer ion conductor.

Segalman et al. added Ag ions to Nafion and PSS membranes to prepare two new material systems: Ag-Nafion and Ag-PSS. Both of these systems are water-processable solid electrolytes [25], which exhibit high ionic conductivity due to their sulfonate groups. The silver ions combined in Ag-Nafion and Ag-PSS undergo directional migration under the action of the temperature gradient and form an ion concentration gradient through the generalized Soret effect, and then generate a thermoelectric voltage. The silver ions also undergo electrochemical reactions on the electrodes, causing silver to enter the system from the anode and deposit on the cathode. Since the sulfonate ions are on the polymer backbone, only the silver ions are movable under the applied temperature gradient, so the Seebeck coefficient is completely the result of the movement of the silver ions. Segalman et al. also analyzed the ion transport under temperature gradient from the chemical reaction entropy and Soret effect, and finally demonstrated a flexible, water-treated solid polymer thermocurrent device. The results show that at 26°C and 50% relative humidity, the maximum ZT of 33%$AgNO_3$-PSS is 0.006, and the maximum ZT of 33%AgOH-Nafion is 0.003. Although its ion mobility is extremely low compared with electrons, this is the first solid polymer thermocurrent device using a single ion conductor, which has good solution processability, flexibility, and air stability.

Liquid polyethylene oxide (PEO) can also be used as an electrolyte for single-ion conductor thermoelectric materials. Crispin et al. added NaOH to PEO [26] to convert the hydroxyl group (-C-OH) at the end of PEO into anionic alkoxide end groups (-CO-Na), where free Na ions are used as charge carriers. This ionic polymer electrolyte can provide a very strong ion Soret effect, with a Seebeck coefficient of up to 10 mV/K and a relatively low thermal conductivity of 0.216 W /(m•K).

In addition to the above-mentioned organic solutions as electrolytes, natural green pollution-free is the development prospect of new thermoelectric materials. Nano cellulose (NFC) is extracted from natural cellulose, fibers, or crystals with a diameter of less than 100 nm and a length of several microns. It has high mechanical strength, flexibility, biodegradability, etc. If the soft nano cellulose is used as the substrate and the conductive polymer with conductive properties to form a composite material, it can be used to prepare a biocompatible energy harvesting device. Crispin et al. added nano cellulose (NFC) flexible additives to PSSNa electrolyte to prepare NFC-PSSNa solid electrolyte [27]. The introduction of nano cellulose makes this solid electrolyte mechanically stronger than pure electrolyte, and its ion Seebeck coefficient is significantly higher than that of pure PSSNa. The preparation of this ionic thermoelectric paper provides new possibilities for the large-scale preparation of ionic thermoelectric materials. Hu Bingliang and others also used natural fibers as ionic thermoelectric electrolytes. They performed a simple chemical treatment on natural wood to make a sub-nano structured cellulose membrane, which was immersed in NaOH solution to make a cellulose-Na mixed electrolyte [28]. The cellulose-Na mixed electrolyte uses Na ions as carriers, and cellulose uses molecular chain arrays to limit ion migration. It has a high selective diffusion ability under thermal gradients. The electrolyte penetrates the cellulose membrane and applies axial temperature gradient, the thermal gradient ratio of the ion conductor (similar to the Seebeck coefficient of thermoelectric) reaches 24 mV/K. This material can be made into a flexible and biocompatible thermoelectric conversion device, which is expected to realize the scale of the thermoelectric conversion device Chemical production.

3.2.3 Ionic liquid quasi-solid ionic gel

Ionic liquid is a kind of salt that consists of only anions and cations and is liquid at room temperature. It has the advantages of low volatility and good thermal stability. Compared with water or common organic electrolyte solutions, ionic liquids have a higher electrochemical window (4 ~ 6 V) can be used to prepare energy storage devices, and is the most promising electrolyte material for thermoelectric capacitors. Ionic liquid electrolyte is mainly composed of ionic liquid with better performance and resin matrix, etc., to make a quasi-solid ionic gel. This gel not only has excellent electrical conductivity and ion transport properties, but the solid gel-like structure also avoids ionic liquids. The loss and environmental pollution problems.

Ouyang's team has made outstanding contributions to the preparation of thermoelectric capacitors based on ionic liquids. Cheng et al. It is mixed into a quasi-solid ionic gel [21], which exhibits ultra-high Seebeck coefficient (up to 26.1 mV/K), high ionic conductivity (6.7 mS/cm), and low thermal conductivity (0.176 W/(m K))), which is the highest Seebeck coefficient detected in electronic and ion thermoelectric materials so far. Water-based polyurethane (WPU) is also a good solid gelling agent. Fang et al. used drop-casting method to mix EMIM: DCA and WPU to prepare a stretchable transparent ionic gel with high thermoelectric properties [29]. The ionic gel exhibits extremely high flexibility and mechanical tensile properties. The tensile performance of the ionic gel containing 40 wt% EMIM: DCA is as high as 156%. It also has good ionic thermoelectric properties. When the relative humidity is 90% With high ionic thermal voltage (34.5 mV/K), high ionic conductivity (8.4mS/cm), and low thermal conductivity (0.23 W/(m K)), the ZT value can reach 1.3 ± 0.2. The advantage of ionic stretchable gel is that it has good transparency and stretchability, which provides a good research idea for the energy supply of flexible wearable devices.

Compared with organic resin gelling agents, the doping of inorganic materials can not only play a good gel effect but also greatly enhance the conductivity of the electrolyte material and further improve its thermoelectric performance. Commonly used inorganic materials are SiO_2 Carbon nanotubes, etc. The Lewis acid group on the surface of SiO_2 nanoparticles interacts with the ionic liquid, which promotes the dissociation of anions and cations, and promotes the formation of more free vacancies, and establishes rapid ion migration channels. Therefore, a small amount of SiO_2 nanoparticles can significantly increase ions. The ionic conductivity of the gel, but the excessive SiO_2 nanoparticles easily form agglomerations to hinder ion migration and transmission, thereby reducing the ionic conductivity. He et al. prepared a quasi-solid ionic gel composed of a variety of ionic liquids and SiO_2 nanoparticles, in which EMIM with 20 wt.% SiO_2 nanoparticles added: DCA ionic gel has a high ion Seebeck coefficient (14.8 mV/K), the excellent ion conductivity ($4.75 \times 10\ S^{-2}$ S/cm) and high power factor (1.04 mW/(m K^2)), the room temperature ion ZT value is as high as 1.47 [30]. In addition, carbon nanotubes also contribute to the improvement of electrical conductivity. Salazar et al. studied ionic liquid electrolytes mixed with multi-walled carbon nanotubes (MWCNT) as an alternative electrolyte for thermoelectrochemical cells [31].

Using liquid polyethylene glycol (PEG) to dope the ionic liquid can also change the type of ionic gel electrolyte. The presence of PEG along the polymer chain hinders the interaction of PVDF/anions, and preferentially binds to cations, thereby promoting the thermal diffusion of cations. EMIM: TFSI/PVDF-HFP fluorinated polymer ion gel is an n-type thermoelectric material, which usually exhibits a negative Seebeck coefficient. After Zhao et al. doped PEG into it, the deviation of the measured value of the Seebeck coefficient changed from -4 mV to +14 mV [32]. The thermal diffusion of ion gel is extremely dependent on the polarity of the polymer matrix. This method can be used to conveniently adjust the type of polymer ion gel electrolyte (p-type or n-type) and provide sufficient energy for ion thermoelectric capacitors. Output voltage and power. Ionic liquid quasi-solid ionic gel electrolyte is newly developed and has broad development and application prospects. It is the most promising electrolyte material for thermoelectric capacitors.

3.2.4 Chemical/physical cross-linking ionic gel

In addition to the method of adding a gelling agent solution, the polymer material can be electrochemically cross-linked with the ionic liquid to prepare a water ionic gel. This ionic gel also has good thermoelectric properties, and its Seebeck coefficient and electrical conductivity are significantly improved, and at the same time, it can also solve the environmental pollution problem caused by the leakage of ionic liquid. Commonly used electrochemical cross-linking methods [33] mainly include thermally initiated cross-linking, light-initiated free radical polymerization, in-situ free radical polymerization, and radiation-induced cross-linking.

Thermally initiated cross-linking is a common chemical crosslinking method. The reaction system includes reactive monomers, thermal initiators, and cross-linking agents. Thermal initiators mainly include azobisisobutyronitrile (AIBN) [34], azobisisoheptaonitrile (ABVN), dibenzoyl peroxide (BPO), etc. The cross-linking agent mainly contains two or more acrylic acid, the most widely used oligomer or monomer of ester group is polyethylene glycol dimethacrylate (PEGDMA). For the first time in the presence of ionic liquid [N2228] Br, Sajid et al. used the free radical initiator AIBN to thermally initiate polymerization/chemical cross-linking PEGDMA to prepare a new chemically cross-linked thermoelectric hydroion gel [35]. Its Seebeck coefficient is 1.38 mV/K, which is more than 10 times higher than that of pure ionic liquid. It also has extremely high ionic conductivity, reaching 74 mS/cm, and the room temperature ZT value reaches 1.02 × 102. In the same year, Sajid et al. replaced the ionic liquid with 1-butyl-3-methylimidazole tetrafluoroborate (BMIM: BF4) [34], and used in-situ free radical polymerization to prepare a thermally initiated cross-linked ionic gel. The results show that the ionic conductivity of the ion gel with 90 wt.% BMIM:BF_4 is also 10 times that of the pure ionic liquid, reaching 49.41 mS/cm. At the same time, the ion Seebeck coefficient is significantly increased, which is 2.35 mV/K, which is a new type of p-type chemically cross-linked ionic gel.

Compared with thermally-initiated cross-linking polymerization, photo-initiated cross-linking does not use initiators and cross-linking agents. It has the advantages of high initiation efficiency, easy control of the polymerization reaction, and mild operating conditions. It is a very effective in-situ polymerization method. Therefore, it is widely used in the preparation of hydroion gel. Liu et al. studied a UV-induced photopolymerization of BMIM: Cl-based ionic gel electrolyte, and finally prepared BMIM: Cl/CS/PHEMA polymer gel electrolyte [36]. This chemically cross-linked ionic gel has Good thermoelectric performance can be used as a cheaper and more scalable way to directly convert waste heat into electrical energy.

In addition, the physical cross-linking method can also be used to prepare hydroion gels. Jang et al. A stretchable, high-performance, flexible and self-healing ternary ion thermoelectric hybrid material composed of sulfonic acid, PAAMPSA) and physical cross-linking agent (phytic acid, PA) [37], it shows very superior ionic thermoelectric properties, Seebeck coefficient is 8.1 mV K-1, ZTi reaches 1.04. Due to the flexibility of the polymer, the ternary hybrid ion gel has good flexibility and ductility, and the dynamic interaction between the three components allows it to heal on its own without external stimulation. It is flexible and wearable. Ionic thermoelectric capacitors provide new features such as stretchability and self-repair.

3.3 Electrode material

Metal electrodes are one of the most widely used electrode materials for ionic thermoelectric capacitors (TEC). In addition to inert metal platinum and gold electrodes, if the carriers in the single-ion conductor electrolyte are silver ions, silver can also be used as the electrode material [25]. This hybrid ion-electronic conductive polymer combines the redox reaction on the electrode and the thermoelectric transport of ions in the system, which can enhance the ionic electrochemical activity of the material. In order to make ionic thermoelectric capacitors have higher performance, TEC electrodes need to have a higher surface area and ion transport channels, thereby having higher conductivity and ion transport rate. Nano-carbon materials such as carbon nanotubes (CNT) and graphene are ideal candidate materials for TEC electrodes.

Carbon nanotubes and graphene and other nano-carbon materials have a unique two-dimensional planar structure and have good performance in ion and electron transport. Compared with silver electrodes, carbon nanotube electrodes have the higher electroactive surface area and higher capacitance than the silver electrode. Crispin et al. used the drop casting method to prepare CNT film on gold film [26]. Due to the high specific surface area of the porous network structure of the nano-electrode, the ionic thermoelectric capacitor exhibits a higher capacitance. Kim et al. deposited polyaniline on nanoporous graphene/carbon nanotube (G/CNT) films by electropolymerization to prepare sandwich electrodes [24], in which carbon nanomaterials provide high surface area, and polyaniline has good redox activity.

This sandwich composite electrode produces 3–7 times higher capacitance than pure CNT. Electrode materials suitable for ionic thermoelectric capacitors are still in the development stage. It is believed that future electrode materials will not only provide higher specific surface area and capacitance but will also develop in flexibility and mechanical stretchability, making them more suitable for small wearable flexible thermoelectric energy harvesting devices.

4. Ionic thermoelectric materials for other energy conversion devices

4.1 Hybrid ion/electronic thermoelectric converter

If the ionic thermoelectric capacitor is combined with an ordinary thermogenic battery, the ion conductive layer is placed on the electronic layer, the electronic unit uses the Seebeck effect under the temperature gradient to obtain heat, and the ion unit can utilize the Soret effect to convert heat into electric energy under temperature fluctuation, this thermoelectric conversion device that combines the electronic Seebeck effect and the ion Soret effect is called a hybrid ion/electronic thermoelectric converter (HTEC). Its working mechanism is shown in **Figure 6** [38]. Due to the Soret effect, the cations in the ion layer will accumulate on the hot side of the ion gel under a temperature gradient, while the additional cations induce electrons to pass through the interface between the electron layers, and neither ions nor electrons can pass through the interface between the two layers. When in equilibrium, the distribution of electrons and ions at the interface is similar to the upper electrochemical layer. The interface in the middle is called the electric double layer, and heat is the driving force for the accumulation of interface charges.

The hybrid ion/electronic thermoelectric converter combines the Seebeck effect of electrons and the Soret effect of ions, which can generate higher thermal voltage and provide new thinking directions for thermoelectric energy conversion. Ouyang et al. coated a layer of PSSH on the top surface of the acid-treated PEDOT: PSS to form a thermoelectric material with a PSSH/A-PEDOT: PSS double-layer structure. Compared with the mixture PEDOT:PSS:PSSH, this double-layer film has more high power factor, reaching 401 mW/(m·K^2) [39]. In addition, similar effects were found in butylamine doped PEDOT:PSS (PSSNa/BA-PEDOT:PSS). Ionic gel can also be used as the ionic unit. Cheng et al. prepared a hybrid ion/electronic

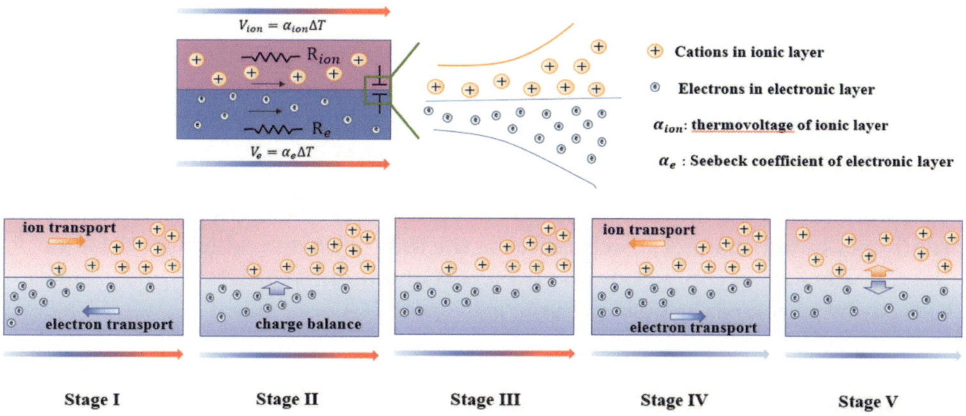

Figure 6.
Schematic illustration of a HTEC with the electronic and ionic layers. R_{ion} and R_e are for the ion and electron resistances under electric field, respectively.

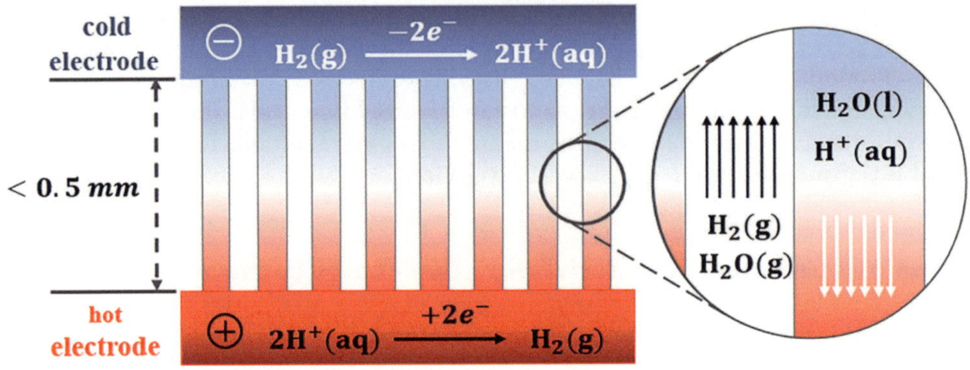

Figure 7.
Working mechanism of gas–liquid two-phase thermoelectric battery.

thermoelectric converter composed of PEDOT:PSS and ionic gel [38]. The ionic layer is composed of 80 wt% 1-ethyl-3-methyl Although the ZT value of the ionic gel is composed of base imidazole dicyandiamide (EMIM: DCA) and 20 wt% PVDF-HFP is not improved, the thermal voltage is significantly increased. The stable voltage can reach 28 μV under a temperature gradient of 0.65 K. The hybrid ion/electron thermoelectric converter not only improves the energy collection but also provides a new research direction for the future development of ionic thermoelectric conversion materials.

4.2 Gas–liquid two-phase thermoelectric battery

Although the thermogenic battery can directly convert waste heat into electric energy by using the oxidation–reduction reaction under the temperature gradient, the check and balance between thermal resistance and conductance limits the improvement of its conversion efficiency. Ma et al. developed a thermoelectric battery involving gas–liquid two-phase, replacing the traditional liquid-phase redox reaction and mass transfer process with gas–liquid two-phase system. They use H_2SO_4 aqueous solution as the electrolyte of the thermoelectric battery [40]. H_2/H^+ redox is a fast reaction involving both H^+ transport and H_2 gas generation. The thermogenic cell reaction process is shown in the **Figure 7**. Among them, H^+ is reduced to H_2 on the hot side, and H_2 is oxidized to H^+ on the cold side. H_2 and H^+ shuttle in a thin separator with ordered gas–liquid interpenetrating channels. The electrode adopts a reversible hydrogen electrode, and the electron transfer resistance is very small. Even if there is only half the volume of the acid solution in the separator, the H^+ conduction is fast enough. The gas-containing electrolyte layer can also suppress the contribution of heat convection to the thermal conductivity of the thermogenic cell. Improve its thermoelectric performance, its power factor is 12.9 $\mu W/(mK^2)$, there is still a lot of room for development.

The unique design of this gas–liquid two-phase thermoelectric battery lies in the orderly gas/liquid channel structure inside the thin separator, which provides an effective way for ion conduction/gas transport and thermal isolation, and realizes the transformation of the power characteristics of TGCs.

5. Summary and prospect

According to the different thermoelectric conversion mechanisms, this paper divides ionic thermoelectric materials into two types: thermoelectric materials for thermoelectrochemical cells (thermocouple effect) and thermoelectric materials for ionic thermoelectric capacitors (Soret effect), and their latest progress in thermoelectric conversion devices is described. In recent years, small thermoelectric conversion devices have gradually developed in the direction of wearable devices, and their thermoelectric performance and conversion efficiency has been greatly improved, and more attention has been paid to the flexibility and stretchability of thermoelectric conversion materials. Compared with traditional electronic thermoelectric materials, the conversion efficiency of ionic thermoelectric materials has been improved, but there is still room for development. In order to achieve higher thermoelectrochemical performance and practical applications, it is necessary to improve the ion mobility of materials and establish ion diffusion channels. Electrode materials should also have a high specific surface area and unobstructed ion diffusion channels to achieve good electron and heat transfer. Nano-carbon materials and PEDOT/PSS and their composite materials are good candidates for ionic thermoelectric electrodes, and they still have broad Development prospects, and at the same time, single-atom doping of platinum, copper, and other metal electrodes is also a good research approach.

Author details

Jianwei Zhang*, Ying Xiao, Bowei Lei, Gengyuan Liang and Wenshu Zhao
Department of Materials Science and Engineering, National University of Defense Technology, Changsha, P.R. China

*Address all correspondence to: jwzhang.nudt@gmail.com

References

[1] Kim SI, Lee KH, Mun HA, et al. Thermoelectrics. Dense dislocation arrays embedded in grain boundaries for high-performance bulk thermoelectrics. Science. 2015;**348**(6230):109-114

[2] Pei Y, Shi X, Lalonde A, et al. Convergence of electronic bands for high performance bulk thermoelectrics. Nature. 2011;**473**(7345):66-69

[3] Joshi G, Lee H, Lan Y, et al. Enhanced thermoelectric figure-of-merit in nanostructured p-type silicon germani-um bulk alloys. Nano Letters. 2008;**8**(12):4670-4674

[4] Bin W, Helong Z, Yu L, et al. Research progress in organic thermoelectric materials. Journal of Nanchang Hangkong University (Social Sciences) 2020, 34(01): 31-42.

[5] Guangbao W. Preparation of N-type Organic Thermoelectric Materials Baesd on Single-Walled Carbon Nanotubes and Organic Small Molecules. Vol. Master. Shandong: Qingdao University of Science and Technology; 2017. p. 83

[6] Shi W, Qu S, Chen H, et al. One-step synthesis and enhanced thermoelectric properties of polymer–quantum dot composite films. Angewandte Chemie. 2018;**130**(27):8169-8174

[7] Quickenden TI, Mua Y. A review of power genera-tion in aqueous thermogalvanic cells. Journal of the Electrochemical Society. 1995; **142**:3985-3994

[8] Jiao N, Abraham TJ, Macfarlane DR, et al. Ionic liquid electrolytes for thermal energy harvesting using a cobalt redox couple. Journal of the Electrochemical Society. 2014;**161**(7): D3061-D3065

[9] Taheri A, Macfarlane DR, Pozo-Gonzalo C, et al. Quasi-solid-state electrolytes for low-grade thermal energy harvesting using a cobalt redox couple. ChemSusChem. 2018;**11**(16): 2788-2796

[10] Taheri A, Macfarlane DR, Pozo-Gonzalo C, et al. Application of a water-soluble cobalt redox couple in free-standing cellulose films for thermal energy harvest-ing. Electrochimica Acta. 2019;**297**:669-675

[11] Laux E, Jeandupeux L, Uhl S, et al. Novel ionic liq-uids for thermoelectric generator devices. Materials Today: Proceedings. 2019;**8**:672-679

[12] Yang P, Liu K, Chen Q, et al. Wearable thermocells based on gel electrolytes for the utilization of body heat. Angewandte Chemie International Edition. 2016;**55**(39):12050-12053

[13] Zhou Y, Liu Y, Buckingham MA, et al. The signifi-cance of supporting electrolyte on poly (vinyl alcohol)–iron(II)/iron(III) solid-state electrolytes for wearable thermo-electrochemical cells. Electrochemistry Communications. 2021;**124**:106938

[14] Wu J, Black JJ, Aldous L. Thermoelectrochemistry using conventional and novel gelled electrolytes in heat-to-current thermocells. Electrochimica Acta. 2017;**225**:482-492

[15] Han C, Qian X, Li Q, et al. Giant thermopower of ionic gelatin near room temperature. Science (American Association for the Advancement of Science). 2020;**368**(6495):1091-1098

[16] Duan J, Yu B, Liu K, et al. P-N conversion in ther-mogalvanic cells induced by thermo-sensitive nanogels for body heat harvesting. Nano Energy. 2019;**57**:473-479

[17] Im H, Moon HG, Lee JS, et al. Flexible thermocells for utilization of

body heat. Nano Research. 2014;**7**(4): 443-452

[18] Romano MS, Gambhir S, Razal JM, et al. Novel carbon materials for thermal energy harvesting. Journal of Thermal Analysis and Calorimetry. 2012;**109**(3): 1229-1235

[19] Romano MS, Li N, Antiohos D, et al. Carbon nano-tube - reduced graphene oxide composites for thermal energy harvesting applications. Advanced Materials. 2013;**25**(45):6602-6606

[20] Kang TJ, Fang S, Kozlov ME, et al. Electrical power from nanotube and graphene electrochemical thermal energy harvesters. Advanced Functional Materials. 2012;**22**(3):477-489

[21] Cheng H, He X, Fan Z, et al. Flexible quasi-solid state ionogels with remarkable seebeck coefficient and high thermoelectric properties. Advanced Energy Materials. 2019;**9**(32):1901085

[22] Chang WB, Fang H, Liu J, et al. Electrochemical effects in thermoelectric polymers. ACS Macro Letters. 2016;**5**(4):455-459

[23] Wang H, Ail U, Gabrielsson R, et al. Ionic seebeck effect in conducting polymers. Advanced Energy Materials. 2015;**5**(11):1500044

[24] Kim SL, Lin HT, Yu C. Thermally chargeable solid-state supercapacitor. Advanced Energy Materials. 2016;**6**(18):1600546

[25] Chang WB, Evans CM, Popere BC, et al. Harvesting waste heat in unipolar ion conducting polymers. ACS Macro Letters. 2016;**5**(1):94-98

[26] Zhao D, Wang H, Khan ZU, et al. Ionic thermoelectric supercapacitor. Energy & Environmental Science. 2016;**9**(4):1450-1457

[27] Jiao F, Naderi A, Zhao D, et al. Ionic thermoelectric paper. Journal of Materials Chemistry A. 2017;**5**(32):16883-16888

[28] Li T, Zhang X, Lacey SD, et al. Cellulose ionic conductors with high differential thermal voltage for low-grade heat harvesting. Nature Materials. 2019;**18**(6):608-613

[29] Fang Y, Cheng H, He H, et al. Stretchable and transparent ionogels with high thermoelectric properties. Advanced Functional Materials. 2020;**30**(51):2004699

[30] He X, Cheng H, Yue S, et al. Quasi-solid state nano-particle/(ionic liquid) gels with significantly high ionic thermoelectric properties. Journal of Materials Chemistry A: Materials for Energy and Sustainability. 2020; **8**(21):10813-10821

[31] Salazar PF, Stephens ST, Kazim AH, et al. En-hanced thermo-electrochemical power using carbon nanotube additives in ionic liquid redox electrolytes. Journal of Materials Chemistry A. 2014;**2**(48):20676-20682

[32] Zhao D, Martinelli A, Willfahrt A, et al. Polymer gels with tunable ionic Seebeck coefficient for ultra-sensitive printed thermopiles. Nature Communications. 2019;**10**(1):1093

[33] Qin X, Yong Y, Shijie R. Research advances on chemical crosslinked gel polymer electrolytes for lithium ion batteries. Journal of Founctional Polymer 2019;32(01): 28-44.

[34] Sajid IH, Aslfattahi N, Mohd Sabri MF, et al. Synthesis and characterization of novel p-type chemically cross-linked ionogels with high ionic seebeck coefficient for low-grade heat harvesting. Electrochimica Acta. 2019;**320**:134575

[35] Sajid IH, Sabri MFM, Said SM, et al. Crosslinked thermoelectric hydro-ionogels: A new class of highly

con-ductive thermoelectric materials. Energy Conversion and Management. 2019;**198**:111813

[36] Liu X, Wen Z, Wu D, et al. Tough BMIMCl-based ionogels exhibiting excellent and adjustable performance in high-temperature supercapacitors. Journal of Materials Chemistry A. 2014;**2**(30):11569

[37] Akbar ZA, Jeon J, Jang S. Intrinsically self-healable, stretchable thermoelectric materials with a large ionic seebeck effect. Energy & Environmental Science. 2020;**13**(9): 2915-2923

[38] Cheng H, Ouyang J. Ultrahigh thermoelectric power generation from both ion diffusion by temperature fluctuation and hole accumulation by temperature gradient. Advanced Energy Materials. 2020;**10**(37):2001633

[39] Guan X, Cheng H, Ouyang J. Significant enhancement in the Seebeck coefficient and power factor of thermoelectric polymers by the Soret effect of polyelec-trolytes. Journal of Materials Chemistry A: Materials for Energy and Sustainability. 2018;**6**(40):19347-19352

[40] Ma H, Wang X, Peng Y, et al. Powerful thermogalvanic cells based on a reversible hydrogen electrode and gas-containing electrolytes. ACS Energy Letters. 2019;**4**(8):1810-1815

Chapter 5

Challenges in Improving Performance of Oxide Thermoelectrics Using Defect Engineering

Jamil Ur Rahman, Gul Rahman and Soonil Lee

Abstract

Oxide thermoelectric materials are considered promising for high-temperature thermoelectric applications in terms of low cost, temperature stability, reversible reaction, and so on. Oxide materials have been intensively studied to suppress the defects and electronic charge carriers for many electronic device applications, but the studies with a high concentration of defects are limited. It desires to improve thermoelectric performance by enhancing its charge transport and lowering its lattice thermal conductivity. For this purpose, here, we modified the stoichiometry of cation and anion vacancies in two different systems to regulate the carrier concentration and explored their thermoelectric properties. Both cation and anion vacancies act as a donor of charge carriers and act as phonon scattering centers, decoupling the electrical conductivity and thermal conductivity.

Keywords: thermoelectrics, nonstoichiometry, defect, phonon scattering, conductive oxide, charge transport

1. Introduction

The demand for alternate energy of fossil fuel becomes a challenging task for the researcher and scientist. A possible strategy for an alternate energy source is thermoelectric (TE) materials, whereby unwanted heat is changed to useful electrical energy with no harmful emissions compared to other traditional power plants [1, 2]. The performance of a TE materials can be evaluated by the figure of merit, $zT = S^2\sigma T/\kappa$, where σ represents the electrical conductivity, S is the Seebeck coefficient, T is the temperature, and κ is the thermal conductivity [3, 4]. The high zT can be obtained with high electrical conductivity, high Seebeck coefficient, and low thermal conductivity. However, the correlations between them are complex and cannot be treated independently [5, 6]. For instance, the increase in the carrier concentration, rises the electrical conductivity, drop the Seebeck coefficient, and rise the electronic thermal conductivity [7]. Therefore, the optimization of TE performance is a challenging task.

To date, many innovative TE materials like Bi_2Te_3, PbTe, $CoSb_3$ XNiSn, and SiGe have been commercially applied because of their high performance as compared to other TE materials [5, 8–12]. However, these materials have restricted applications due to their high price, instability in oxidizing atmospheres, and the most important toxicity [8, 13–15]. Therefore, many studies on alternative TE materials with low cost, high efficiency, and environmentally friendly characteristics have been explored [16–18]. In this relay, oxide materials have been considered as the best alternative, such as layered cobalt oxide ($NaCo_2O_4$), and strontium titanium oxide ($SrTiO_3$) due to their low price, thermal stability, and eco-friendly compatibility [19, 20]. For TE applications, n-type and p-type materials should be coupled. So it is important to develop high performance both n-type and p-type materials.

Layered structure p-type $NaCo_2O_4$ has been explored and is considered as one of the candidate materials for TE applications [16, 21]. This layered structure contains two layers CoO_2 and Na. The CoO_2 layers play the role of electron source which results in high electrical conductivity, while the layer of Na ions is sandwiched among nearby CoO_2 layers which decreases the thermal conductivity along the stacking direction [22, 23]. Various studies propose that such layered structured CoO_2 exhibits low thermal conductivity with metallic-like electrical conductivity, which is very attractive for TE applications [16, 24].

Similar to highly studied $NaCoO_2$ TE material, $LiNbO_2$ has a layered configuration in which the NbO_6 trigonal-prismatic layers and Li planes are stacked, as shown in the schematic **Figure 1(a)**. This similarity proposes that $LiNbO_2$ could be a new promising TE material. $LiNbO_2$ is a sub-oxide of the main ($LiNbO_3$), and its Li-intercalated structure $Li_{1-x}NbO_2$ was first explored as a promising superconductor [25–27]. It has also been considered as a potential candidate material for numerous technological applications [28–31]. The removal of Li atoms provide additional holes to the valence band, made up of Nb d_Z^2 states, which affects the oxidation of the Nb-atoms and raises the density of states (DOS) at the Fermi level (just like when holes are introduced into $NaCoO_2$). These Li-vacancies increase the carrier concentration, which will raise the electrical conductivity. Moreover, the intrinsic defects in $Li_{1-x}NbO_2$ would act as a scattering center for thermal conductivity. So, defected $Li_{1-x}NbO_2$ is estimated to be extremely favorable for TE applications because of higher electrical and lower thermal conductivities.

Stoichiometric $SrTiO_3$ has a cubic perovskite structure, where oxygen anions form an octahedron with one Ti^{4+} atom lying at the center as shown in the schematic

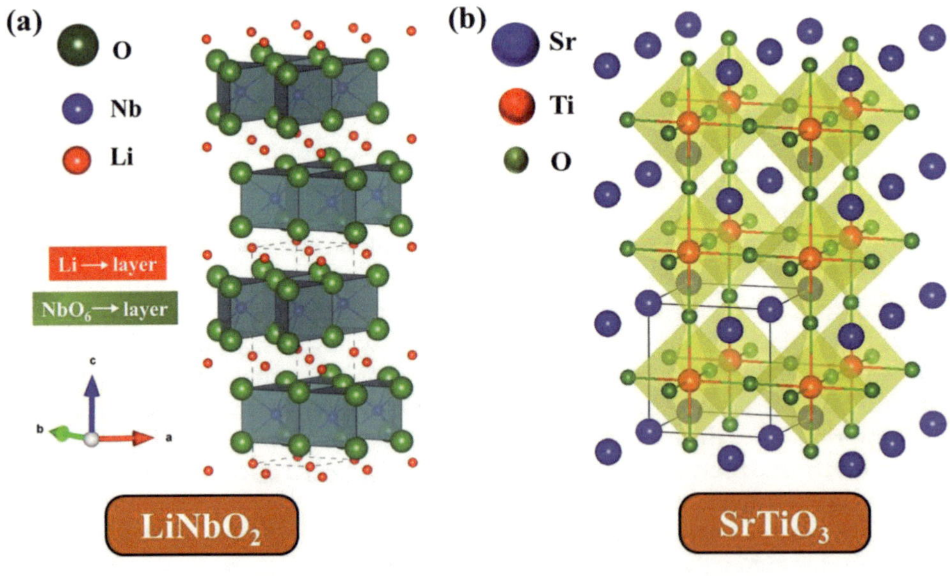

Figure 1.
Schematic illustration of (a) $LiNbO_2$ (b) $SrTiO_3$.

Figure 1(b). $SrTiO_3$ have been studied widely and is considered as one of the favorite *n*-type TE materials due to its high absolute Seebeck coefficient [32, 33]. Virgin $SrTiO_3$ is considered to be an insulator and with a bandgap of 3.25 eV [34]. To boost the power factor (PF), only the electrical conductivity needs to be increased through appropriate doping on A-site and/or B-site [19, 35–39]. The role of the oxygen vacancy, which also acts as electron dopants, is very important for TE materials. Besides, creating oxygen vacancies offer a chance to decrease thermal conductivity through phonon scattering without remarkably affecting electrical conductivity [40–42]. The cationic nonstoichiometry and controlling oxygen partial pressure in $SrTiO_3$ can produce cation and oxygen vacancies, which play an important role in the TE performance. To understand the defect chemistry and its consequence on the TE performance, pure $SrTiO_3$ should be considered.

Considering the importance of defects in oxide materials as discussed above, it is important to study cation and anion defect engineering in oxide materials. Here in this chapter we have considered $LiNbO_2$ (*p*-type) and $SrTiO_3$ (*n*-type) with different vacancy concentrations and explore the experimental observations and correlate them with density functional theory (DFT). We elucidate that the defect engineering which may provide a new track for enhancing the TE performance.

2. Cation defect engineering

2.1 Experimental and computational approaches

2.1.1 Preparation of $Li_{1-x}NbO_2$ compounds

Nonstiochiometric $Li_{1-x}NbO_2$ (x = 0–0.6) compounds were prepared by conventional solid-state reaction using commercially available Li_2CO_3, NbO, and Nb_2O_5 with purity level more than 99.99%. First, $Li_{3-y}NbO_4$ was prepared by mixing Li_2CO_3 and Nb_2O_5 (y = 0, 0.015, 0.3, 0.6, 0.9, 1.2, and 1.8) at 1173 K in air for 50 hrs. Next, $Li_{1-x}NbO_2$ (x = 0–0.6) were prepared by mixing $Li_{3-y}NbO_4$ (y = 0, 0.015, 0.3, 0.6, 0.9, 1.2, and 1.8) for corresponding x = (0, 0.05, 0.1, 0.2, 0.4 and 0.6), and NbO in a ratio of 1: 2 for 24 hrs in ethanol. The dried slurries were crushed, sieved, and consulidated by spark plasma sintering (SPS) at 1323 K under 50 MPa for 15 min [43].

2.1.2 Theoretical calculations of $Li_{1-x}NbO_2$

The electronic structures of $Li_{1-x}NbO_2$ were explored using DFT with the local density approximations [44]. The lattice constants were calculated and all the atomic sites were relaxed till the forces were met to less than 0.01 eV/Å. For Li-vacancies computations various supercells (1x1x1, 1x1x2, 2x2x1, and 2x2x2) were considered. For the TE calculations, the BolzTraP package based on the Boltzmann transport theory and a constant relaxation time were applied [45, 46]. For band structures calculations QE code and the energy dispersion relation $E(n,k)$ as a function of band index n and wave vector k were used.

2.1.3 Characterization of $Li_{1-x}NbO_2$

X-ray diffraction (XRD) analysis was performed by Rigaku D/MAX-2500/PC with Cu Kα emission. The pictographs of the samples were observed by using the scanning electron microscope (SEM, Verios 460 L, FEI). Rectangular specimens were cut for the measurements of the electrical properties using (ZEM-3, ULVAC-RIKO). Temperature-dependent charge transport properties were measured using

HT-Hall, Toyo Corporation, ResiTest 8400. Circular discs were used for thermal diffusivity measurement (DLF-1300, TA instrument). All samples have ≥95% of the theoretical density.

2.2 Results and discussion

Figure 2(a) displays the XRD patterns of all samples. The main peaks of the compounds were indexed according to the $LiNbO_2$ hexagonal structure which can be regarded as alternatively arranged close-packed Li-layers inserted among the two O-Nb-O slabs along the c-axis [29]. In addition to the main peaks, all samples showed small impurity peaks of $LiNbO_3$ and NbO_2. $LiNbO_3$ peaks are expected due to moderate PO_2 level during the consolidation process, which changes the oxidation state from Nb^{3+} to Nb^{5+}. The additional NbO_2 peaks can be described by the following defect reaction (1).

$$2NbO + Li_3NbO_4 \rightarrow 3LiNbO_2 \tag{1}$$

At lower Li-vacancy concentrations, the Li atoms to combine with Nb^{3+} ($[1/3Nb^{5+}+2/3Nb^{2+}]^{3+}$) make $LiNbO_2$. But, at high Li-vacancy concentrations, there are insufficient Li-atoms to react with Nb^{3+} ($[1/3Nb^{5+}+2/3Nb^{2+}]^{3+}$) atoms to form $LiNbO_2$ which turn into the NbO_2 phase. Furthermore, at lower oxygen partial pressure the Nb^{5+} in the Li_3NbO_4 compound is thermally reduced to Nb^{4+} which is the consequence of the NbO_2 phase. Additionally, the Li-vacancies lead to an increase in the repulsive force among the two adjacent oxygen layers which increases the *c*-lattice of the unit cell. In the meantime, the *a* lattice decreases due to the shrinkage of Nb-O bonds in NbO_6 octahedra. The experimental lattice constant shown in the inset **Figure 2(a)** is close to reported work and with our DFT results [47, 48].

Figure 2(b)-(f) displays the fractured cross-section of $Li_{1-x}NbO_2$ samples. The dense of all the samples support the high relative densities larger than 95%. All grains are homogeneously distributed, no obvious segregations, and are randomly oriented. Moreover, the Li-vacancy concentrations have no significant effects on the size and shape of the grains.

Figure 3(a) shows the temperature-dependent electrical conductivities of $Li_{1-x}NbO_2$ samples. At first, the electrical conductivities of the samples increased with Li-vacancy concentrations, suggesting that more holes are created, as shown in the defect reaction [2]. The decreasing trends in the electrical conductivity

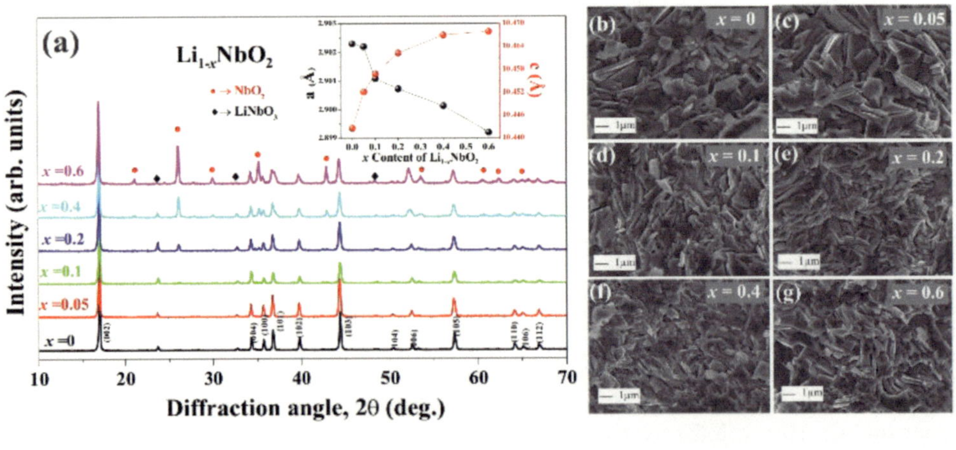

Figure 2.
(a) X-ray diffraction patterns and (b-g) microstructure for $Li_{1-x}NbO_2$ samples [43].

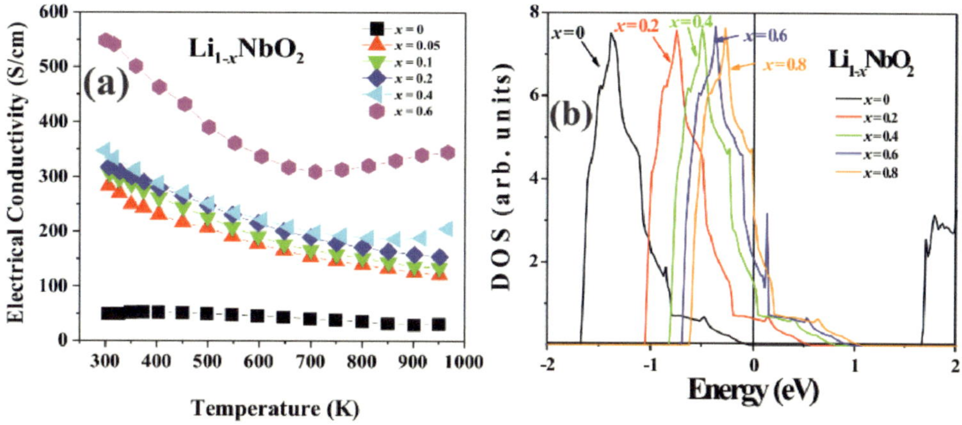

Figure 3.
(a) Temperature dependence of electrical conductivity and (b) calculated electronic density of states of $Li_{1-x}NbO_2$ (x = 0, 0.2, 0.4, 0.6, and 0.6) [43].

measurements with temperature suggesting metallic behavior. However, for higher Li-vacancies concentrations, i.e., $x \geq 0.4$, and at high temperatures the metallic conduction changed to semiconducting-like behavior. For high Li-vacancy concentrations, this transition point is moved to lower temperatures, suggesting that hole creation is inhibited by electrons formed by the replacement of Nb with Li atoms. Above this transition point, the electrical conductivity is governed by electrons which can be described by the defect reactions displayed in Eqs. (3)–(5).

$$Li_2O + Nb_2O_3 \rightarrow 2V_{Li}^{'} + 2h^{\cdot} + 2Nb_{Nb}^{X} + 4O_O^X \quad (2)$$

$$Li_2O + Nb_2O_3 \rightarrow V_{Li}^{'} + Nb_{Nb}^{\cdot\cdot} + Li_{Li}^{X} + Nb_{Nb}^{X} + 4O_O^X \quad (3)$$

$$Li_2O + Nb_2O_3 \rightarrow Nb_{Li}^{\cdot\cdot} + V_{Li}^{'} + e' + Nb_{Nb}^{X} + 4O_O^X \quad (4)$$

$$Li_2O + Nb_2O_3 \rightarrow Nb_{Li}^{\cdot\cdot} + 2e' + Li_{Li}^{X} + Nb_{Nb}^{X} + 4O_O^X \quad (5)$$

To deeply understand the above experimental results DFT calculations were also employed and the electronic structures were calculated. The calculated electronic structure suggests that virgin $LiNbO_2$ has a band-gap of 1.65 eV. It should be noted that DFT-LDA generally miscalculates the band-gap. But, the calculated band-gap in our work is consistent with the reported works [45]. **Figure 3(b)** summarizes the electronic DOS for various holes concentrations. It suggests that virgin $LiNbO_2$ is a semiconductor and suggesting a metallic behavior for holes incorporated samples. Besides, the DOS at the Fermi energy also increases, suggesting that the electrical conductivity of $Li_{1-x}NbO_2$ should increase with increasing holes concentrations.

Figure 4(a) shows the temperature-dependent Seebeck coefficients (*S*) for all samples. It can be seen that both the calculated and experimentally observed *S* are very close. The positive sign indicates that holes are the majority carrier which can be connected to the deviation from stoichiometry in the Li- sublattice [49]. The S values of $x \leq 0.1$ samples increase with temperature, showing a degenerately doped semiconductor. Sample with $x = 0.2$ and at high temperatures the increasing trend in the S values is low suggesting that the hole generation is suppressed by an

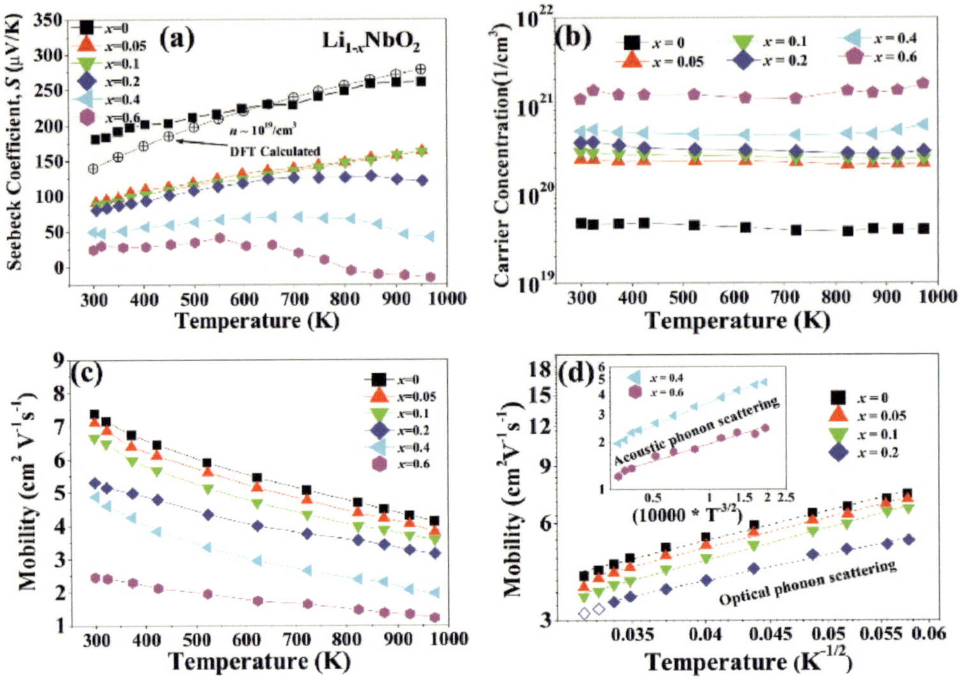

Figure 4.
Temperature dependence of (a) Seebeck coefficient, (b) carrier concentration, (c) carrier mobility, and (d) normalized mobility as a function of $T^{r-1}\left(r=\frac{1}{2}\text{ and }-\frac{1}{2}\right)$, of $Li_{1-x}NbO_2$ *[43].*

electron. Additional increase in the Li-vacancy concentrations ($x > 0.2$) and at high temperatures, the S values starts decreasing, and finally, a changeover is observed. This changeover from p-type to n-type could be due to cation disorder which is very similar to the σ-T as displayed in **Figure 3(a)**. The behavior could be clearly understood by the defect reaction (2)– (5).

Figure 4(b)-(d) shows the temperature-dependent Hall measurements for all $Li_{1-x}NbO_2$ samples. The obtained by the Hall measurements agreed with the electrical conductivity and Seebeck coefficient data. It can be seen clearly that the carrier concentration increases with the creations of Li-vacancies $\left(2V_{Li}^{'} \rightarrow 2h^{\cdot}\right)$. Samples with lower Li-vacancy concentrations ($x \leq 0.2$) were not influenced by temperature, suggesting degenerately doped behavior. However, high Li-vacancy concentrations ($x \geq 0.4$), and at high temperatures show a slight rise in the carrier concentration. This tendency is similar to the Seebeck coefficients as shown in **Figure 4(a)**. In the case of a degenerately doped p-type semiconductor, the Fermi level lies below the valence band maxima (VBM) and the carrier concentration is temperature independent of up to the intrinsic-extrinsic transition temperature [50]. Therefore, it is expected that $Li_{1-x}NbO_2$ is a heavily doped semiconductor. **Figure 4(c)** shows the temperature dependence mobility of nonstoichiometric $Li_{1-x}NbO_2$ compound and the carrier mobility decreases with increasing temperature. Together with the Hall measurements and the defect reactions (2)– (5), it is expected that the tendency in electrical conductivity is directed by mobility in the region holes are dominant and by carrier concentration in the region where electrons are dominant ($x \geq 0.4$ and at high temperature). All samples show negative temperature-dependence carrier mobility, resulting from the phonon scattering ($\mu \propto T^{r-1}$), where $r=-\frac{1}{2},\frac{1}{2}$, and $\frac{3}{2}$ signify acoustic, optical, and ionized impurity phonon scattering, respectively [51–53]. To know the scattering mechanism in $Li_{1-x}NbO_2$, mobilities were re-plotted as a function of T^{r-1} and shown in **Figure 4(d)**. The samples $x \leq 0.2$ shown a linear correlation with $T^{-1/2}$, suggesting that mobility is dominantly encountered by optical phonon scattering. However, samples with $x \geq 0.4$, as shown in **Figure 4(d)** inset, displayed a linear relationship with $T^{-3/2}$, suggesting that the mobility is encountered by acoustic phonon scattering. The transition mechanism is not clear, however, it could be due to the dominant defect change with Li-vacancies, corresponding to Equations (2) and (3), to Nb replacements for Li sites, corresponding to Eqs. (4) and (5).

Figure 5(a) represents the temperature-dependent total thermal conductivities (κ_{tot}) of $Li_{1-x}NbO_2$ samples. It can be seen that total thermal conductivity is decreased significantly with increasing Li-vacancies. The Wiedemann-Franz relationship was used to estimate the electronic and lattice thermal conductivity as shown in the inset of **Figure 5(a)** [54]. It is found that the main influence on the total thermal conductivity originates from lattice vibrations and decreased substantially with Li-vacancies. This suggests that Li-vacancies act as a scattering center for phonons. However, for higher vacancy concentrations ($x \geq 0.2$), the thermal conductivity increases to some extent. This increase could be due to a change in the scattering mechanism from optical phonon to acoustic phonon by $Nb_{Li}^{\cdot\cdot}$ contribution as described above. This mechanism is schematically illustrated in **Figure 5(b)-(d)**. Therefore, it could be suggested that samples with low Li-vacancies concentrations may have localized

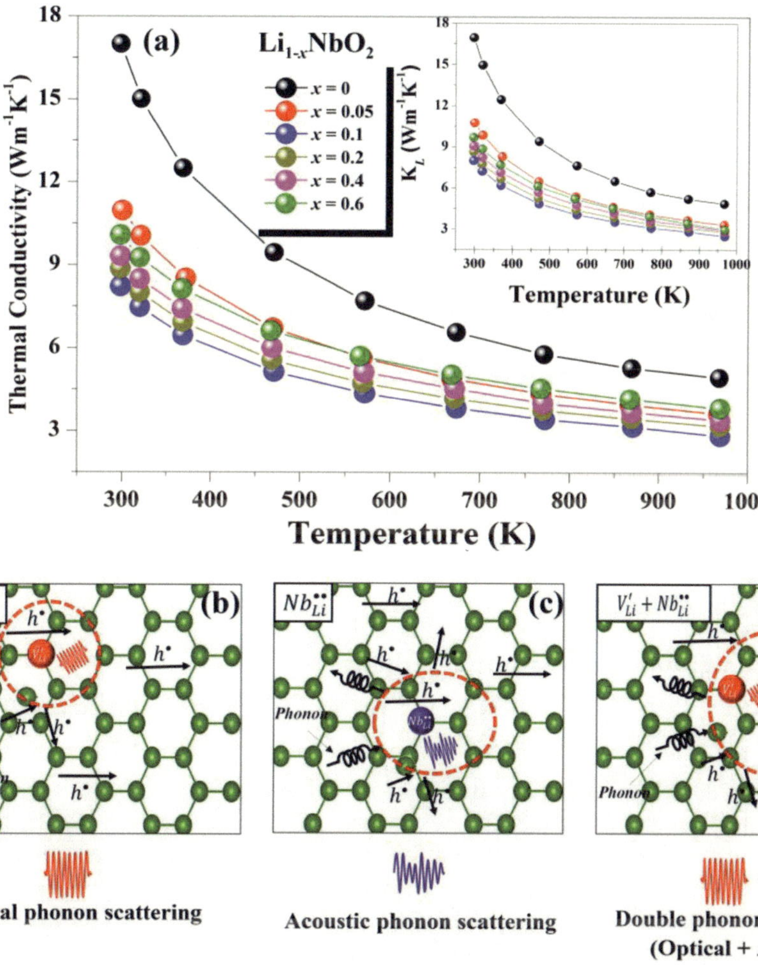

Figure 5.
Temperature-dependent (a) total thermal conductivities for $Li_{1-x}NbO_2$ sample and (b-d) schematic illustration of phonon scattering [43].

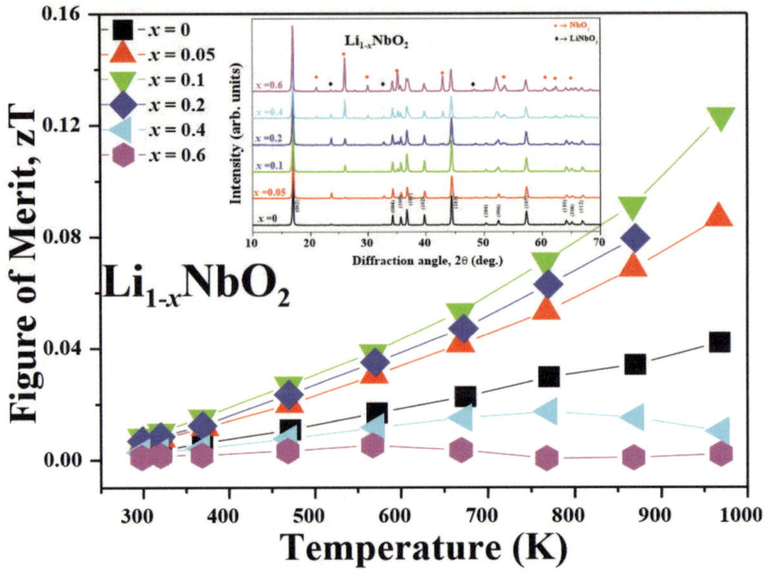

Figure 6.
Temperature dependence of the dimensionless figure of merit of $Li_{1-x}NbO_2$ samples [43].

double phonon (optical and acoustic phonons), which may play a role as a phonon scattering center, showing a substantial decrease of lattice thermal conductivity. However, with higher Li-vacancies concentrations, the lattice vibration is scattered by a single phonon type of acoustic phonon. Also, the increase in the thermal conductivity could be due to secondary phases [55] as detected in XRD (see **Figure 2(a)**).

The figure-of-merit (zT) for all $Li_{1-x}NbO_2$ samples is presented in **Figure 6**. A substantial improvement in zT is observed in the whole temperature range and reaches a maximum value of 0.125 at 970 K, which is around ~220% (3-times) higher compared to the virgin $LiNbO_2$ sample. The observed tendency suggests that zT could be higher at higher temperatures. The samples after high-temperature measurement were rechecked by XRD and found that all $Li_{1-x}NbO_2$ samples are highly stable as shown in the inset of **Figure 6**. This confirms that all samples are stable and can be utilized as a new promising material for high-temperature TE applications.

3. Anion defect engineering

3.1 Experimental and computational methods

3.1.1 Preparation of $SrTiO_{3-\delta}$

Pristine $SrTiO_3$ samples were synthesized by using conventional solid-state reaction techniques, using TiO_2 and $SrCO_3$ with a purity level higher than 99.9%. The stoichiometric powders were ball-milled for 24 hr. Next, the ball-milled powders were calcined at 1373 K for 3 hrs. The powders were then sieved and pressed into thicknesses of 3 mm. The pressed pellets were sintered at 1573 K for 30 hrs under the air atmosphere. To create an Anion defect in $SrTiO_3$ (Oxygen vacancies), the samples were annealed by allowing 1%H_2/Ar, 5%H_2/Ar, 10%H_2/Ar, and 20%H_2/Ar gases at 1573 K for 30 hrs, and the samples were designated with the prefix "1HAr, 5HAr, 10HAr, and 20HAr", respectively.

3.1.2 Theoretical calculations of $SrTiO_{3-\delta}$

To understand the anion defect engineering in $SrTiO_3$, DFT with the local density approximations were applied and correlated with experimental work [44, 56]. To properly pronounce the electronic band structures of pristine and O-vacancies in $SrTiO_3$, we also considered the LDA + U methodology by selecting the effective on-site Coulomb modification (U = 5.0 eV) is applied to d-orbital electrons in Ti-atom in agreement with the reported works [57, 58]. To compute the oxygen vacancy in $SrTiO_3$, a 2 × 2 × 2 supercell was considered and the BoltzTraP program was used for TE properties [45, 46].

3.1.3 Charascterization of $SrTiO_{3-\delta}$

For $SrTiO_{3-\delta}$ characterizations, the same techniques were followed as described in section 2.1.3.

3.2 Results and discussion

Figure 7 shows the XRD pattern of the reduced $SrTiO_{3-\delta}$ samples. The XRD patterns of all reduced samples exhibit a single-phase perovskite cubic structure. The lattice constant of the reduced samples was refined through Rietveld refinements. The lattice constant increased with decreasing P_{O_2} levels. The variations in the lattice constant are similar to the reported values [36, 59]. The variations in the lattice constant indicate that during the annealing process, the oxygen in $SrTiO_3$ sublattice combine with H_2 which increases the O-vacancies among the cations, and thus the coulombic repulsion force of cations increases. This leads to a rise in the lattice constant as presented in **Table 1**.

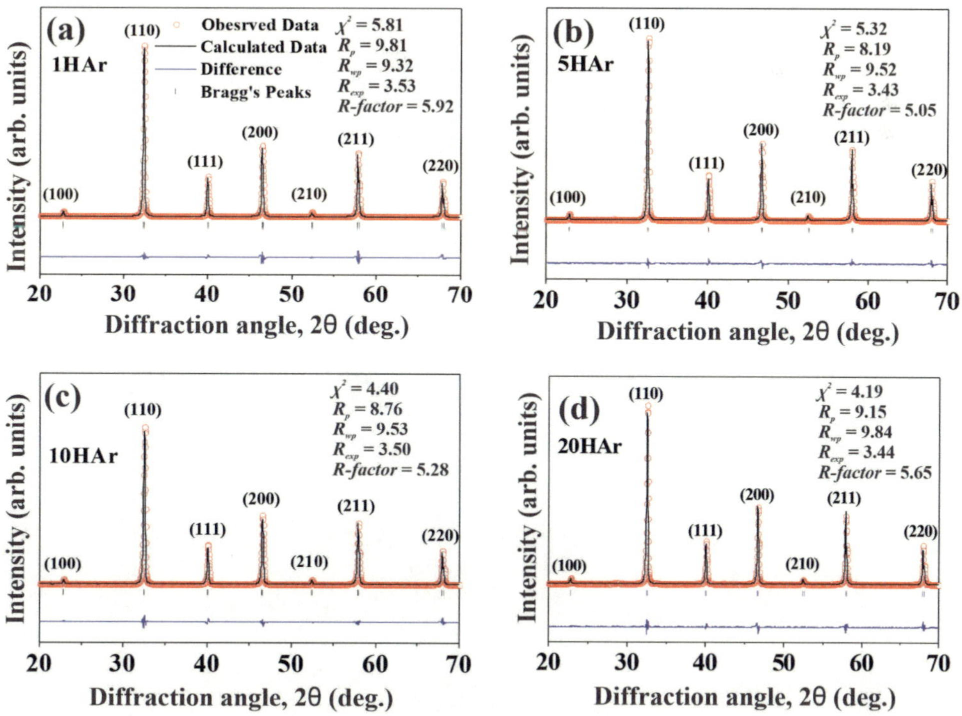

Figure 7.
Refined XRD patterns of oxygen-deficient $SrTiO_{3-\delta}$ ceramics annealed at 1573 K for 30 hrs in (a) 1%H2/Ar (1HAr), (b) 5%H2/Ar (5HAr), (c) 10%H2/Ar (10HAr), and (d) 20%H2/Ar (20HAr) atmospheres [60].

Sample	Lattice parameter (Å)	Carrier concentration ($\times 10^{19}$ cm^{-3})	Oxygen vacancies ($\times 10^{19}$ cm^{-3})	Mobility (cm^{2}V^{-1} s^{-1})	Estimated P_{O_2} (atm)
[1HAr]	3.8991	0.106	0.0530	6.316	2.24×10^{-11}
[5HAr]	3.8999	0.198	0.099	5.373	7.78×10^{-15}
[10HAr]	3.9012	5.24	2.62	4.885	2.23×10^{-23}
[20HAr]	3.9047	10.660	5.33	3.694	3.17×10^{-25}

Table 1.
The carrier concentration, the lattice constant, oxygen vacancies, mobility, and estimated P_{O_2} of $SrTiO_{3-\delta}$ samples [60].

Figure 8 shows the microstructure of thermally etched $SrTiO_{3-\delta}$ samples in high vacuum conditions. It shows that all samples have similar microstructures, well densified, and support the measured densities>95%. There is no substantial variance in the size and shape of the grains.

Figure 9 displays the high-resolution transmission electron microscopy (HRTEM) of oxygen-deficient $SrTiO_{3-\delta}$ samples. It is interesting to see that the sample reduced under different PO_2 levels shows different reduction levels. In the case of low oxygen-deficient samples, we could not see a clear defective area. However, for the high oxygen-deficient samples, we identify two different kinds of defected areas labeled by A1 and A2 in **Figure 9(c,d)**. The area labeled by A1 denotes the local lattice defect which is due to high oxygen vacancy in the lattice and area A2 shows a different lattice structure from the matrix. It is expected that the

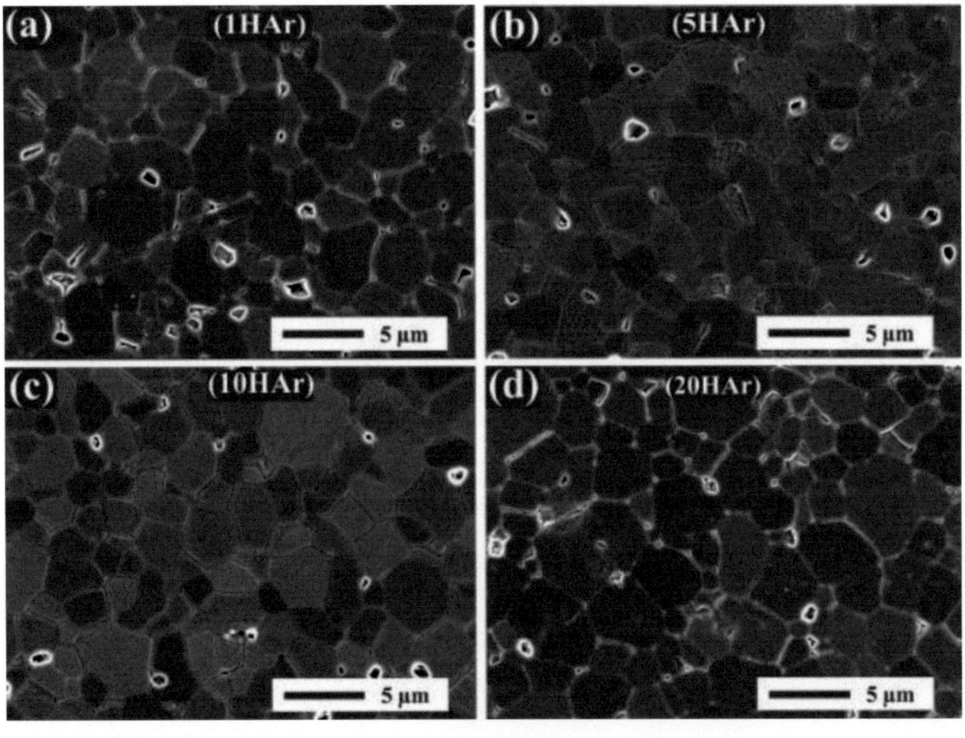

Figure 8.
$SrTiO_{3-\delta}$ ceramics annealed at 1573 K for 30 hrs in (a) 1%H2/Ar (1HAr), (b) 5%H2/Ar (5HAr), (c) 10%H2/Ar (10HAr), and (d) 20%H2/Ar (20HAr) atmospheres [60].

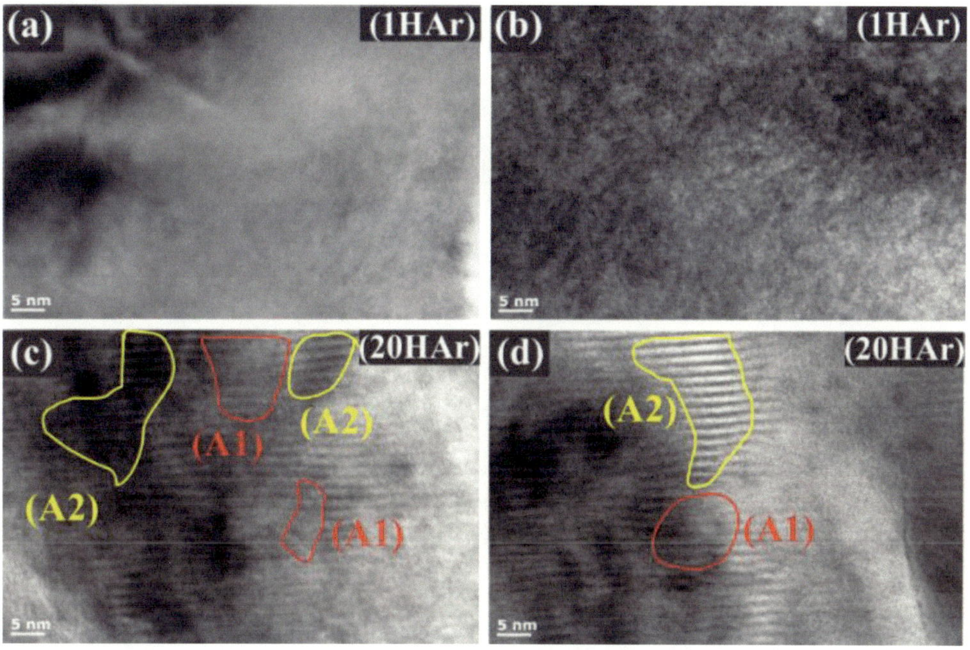

Figure 9.
High-resolution transmission electron microscopy for 1HAr (a, b) and 20HAr (c, d) $SrTiO_{3-\delta}$ [60].

area A2 is probably related to the Ruddlesden-Popper phases [$SrO\cdot(SrTiO_3)_n$] [61], which could be the reason for the metallic conduction in highly reduced samples.

Annealing under a reducing atmosphere at ambient temperature leads to different kinds of defects in materials. Oxygen vacancies could be one of the main defects in the $SrTiO_3$ due to the low formation-energy compared to other atoms in the lattice. Below Eq. (6) shows in the Kröger-Vink notation [62].

$$O_O^x \leftrightarrow \frac{1}{2}O_2 + V_O^{\cdot\cdot} + 2e' \tag{6}$$

where $V_O^{\cdot\cdot}$ denotes the doubly ionized oxygen vacancy and O_O^x denotes the neutral oxygen that exists on its lattice position. The quasi-free electrons generated by oxygen vacancies was determined using the Hall measurements and then using the free electrons, the oxygen partial pressure PO_2 was estimated using Eq.(7) (See **Table 1**) [63].

$$P_{O_2} = \left(\frac{M 2^{\frac{1}{3}} K_0^{\frac{1}{3}} \exp\left(\frac{-\Delta h_{Re}}{3k_B T} \right)}{n_e} \right)^6 \tag{7}$$

where Δh_{Re} represents the enthalpy for reduction, M is the number of oxygen atoms in unit volume (cm^{-3}), k_B is the Boltzmann's constant, K_0 is a constant including an entropy term, n_e is nominal carrier concentration, and T is the annealing temperature.

The observed carrier concentrations in our samples are in the range of 10^{18} to 10^{20}/cm^3, which is a typical range for Mott transition, i.e., insulator–metal transition. This transition can be calculated from the Mott criterion, $n_e^{1/3}a_0 \sim 0.25$, where a_0 is the Bohr radius related to the carrier and n_e which represents the electronic carrier concentration [64]. In the case of 1HAr and 5HAr samples, the calculated Mott transition is on the boundary of Mott transitions, and 10HAr and 20HAr samples are metallic-like, which supports the temperature-dependency of electrical conductivity.

The temperature-dependent electrical conductivity and Seebeck coefficient of reduced $SrTiO_{3-\delta}$ samples are represented in **Figure 10(a)** and **(b)**. As abovementioned that each O-vacancy produces two electrons through charge neutrality conditions, $n_e \approx 2\left[V_O^{\bullet\bullet}\right]$, and henceforth the carrier concentrations would increase, causing an enhance in the electrical conductivity [63]. Moreover, as shown in **Figure 10(a)** that the electrical conductivities for low reduced samples i.e., 1HAr and 5HAr-annealed are low (0.5 to 2 S/cm, correspondingly) in the entire temperature with semiconducting-like. However, in the case of highly reduced samples (increasing O-vacancies (δ) as shown in **Table 1**) i.e., 10HAr and 20HAr samples, the electrical conductivities are relatively higher (41 and 63 S/cm, correspondingly). This increase of electrical conductivity evidently suggests electron doping due to oxygen vacancy, which is analogous to donor-doped $SrTiO_3$ [65]. This would fill the *n*-type conduction band (CB), which evident by the negative sign of the Seebeck coefficient as shown in **Figure 10b**. Additionally, the increase in *n*-type charge carrier from 1HAr to 20HAr is consistent with the electrical conductivity and Seebeck coefficient data.

DFT calculations were also used to understand the effect of O-vacancies in $SrTiO_3$. The electronic band structures of virgin and O-deficient $SrTiO_3$ samples were calculated using DFT + U, and the results are presented in **Figure 10(c)**. Our DFT + U calculated electronic band-gap results suggest that of virgin $SrTiO_3$ at

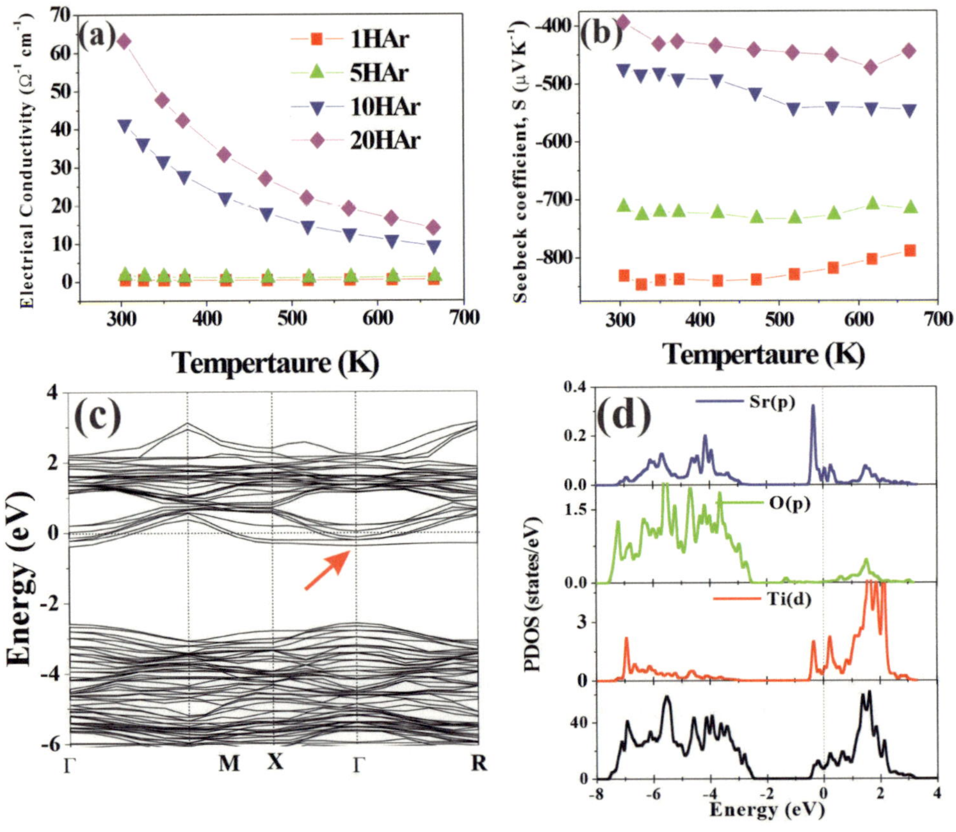

Figure 10.
(a) Temperature-dependent electrical conductivity and (b) Seebeck coefficient of O-deficient $SrTiO_{3-\delta}$, (c) band structure, and (d) total and projected density of states of $SrTiO_{3-\delta}$ [60].

Γ-point is 2.31 eV (1.93 eV), which is close reported work [34, 66]. However, in the case of oxygen-deficient $SrTiO_3$, the electronic band-gap is reduced to 2.18 eV at Γ-point. The band structure calculations show a band below the conduction band, and an electron pocket which can be seen at the Γ- point. Such an electronic pocket suggests electrons in the conduction band, which is largely formed by the Ti-*d* electrons. This electron pocket not only decreases the bandgap but also raises the carrier concentrations that can additionally improve the electrical conductivity of $SrTiO_{3-\delta}$. The projected density of states in **Figure 10(d)** shows that the electron pockets are primarily contributed by the Ti-*d* and Sr-*p* electrons nearby the oxygen vacancy. We also found that at a large O-vacancy concentration, an insulator–metal transition occurs. In addition to this, the charge transfer study also suggests that O-vacancies change the conduction mechanism from nondegenerate to degenerate. We also calculated the room temperature DOS effective mass $\left(m_d^*\right)$ using nondegenerate and degenerate models Eqs. (8) and (9).

$$S = -\frac{k_B}{e}\left[\ln\left(N_c / n_e\right) + A\right] \tag{8}$$

$$S = \frac{8\pi k_B^2}{3eh^2}\left(\frac{\pi}{3n}\right)^{\frac{2}{3}} m_d^* T \tag{9}$$

where k_B represents the Boltzmann constant, h is the Planck constant, n is the carrier concentration, $Nc(\mathrm{T}) = 2\cdot\left[\left(2\pi k_B T m^*\right)/h^2\right]^{3/2}$ is the effective density of states in the conduction band, A is the scattering factor, usually ranges between 0 and 4. By fitting Eq. (8) as shown in **Figure 11a** with A = 1, 2, and 3, it can be seen that the effective mass for low O-vacancies samples suggests a nondegenerate semiconductor. However, the Pisarenko relation fitting Eq. 9 suggests that high O-vacancies samples have degenerate semiconducting like behavior. Furthermore, the DOS effective mass rises with increasing O-vacancies as shown in **Figure 11b**. This increase from 0.43 m_o to 4.4 m_o could be a reason for Hall mobilities reduction as shown in **Table 1**.

Figure 12(a) represents the temperature-dependent thermal conductivity of $SrTiO_{3-\delta}$ samples. This decreasing tendency suggests that O-vacancies act as phonon scattering centers despite their electrical conductivity increased as shown in **Figure 10(a)**. Using Wiedemann-Franz law relation with Lorentz number of

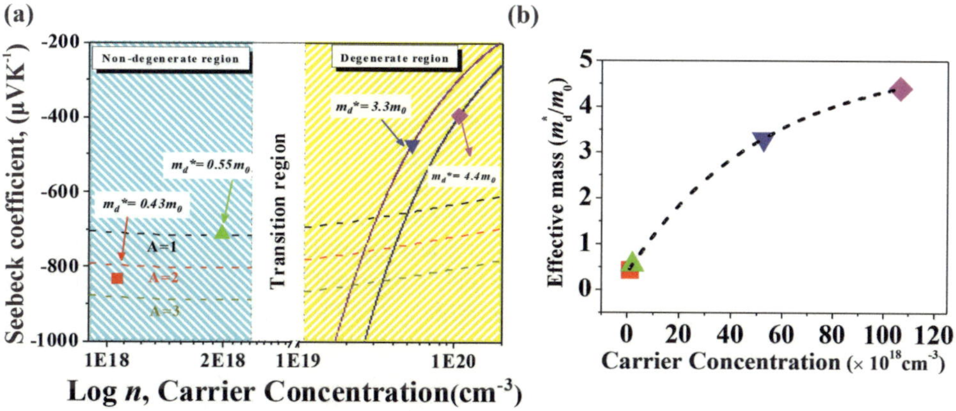

Figure 11.
(a) Seebeck coefficients as a function of carrier concentration: The solid lines represent the estimated Seebeck coefficients using a degenerate semiconducting model and dashed lines for a nondegenerate semiconducting model. (b) DOS effective mass m_d^* resulting from Seebeck coefficient and carrier concentration [60].

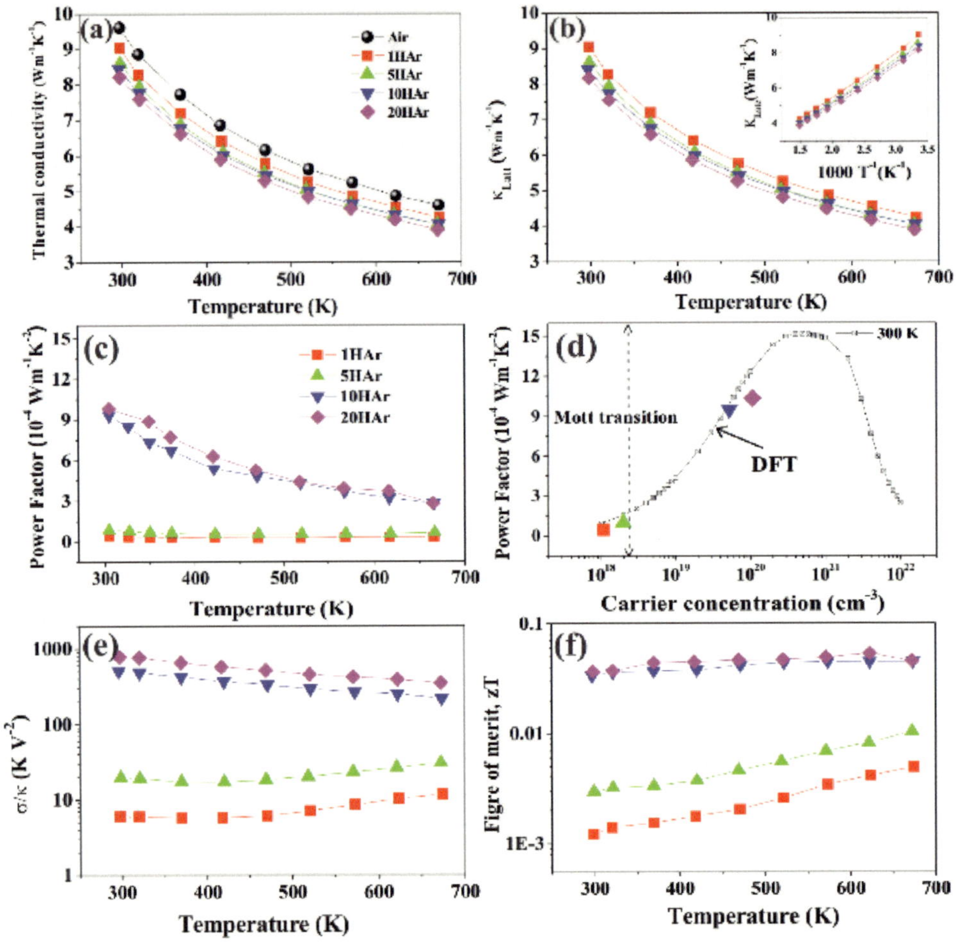

Figure 12.
Temperature-dependent (a) total thermal conductivity, (b) lattice thermal conductivity, (c) power factor, (d) DFT calculated powder factor for various carriers concentrations, (e) temperature-dependent ratio of the electrical to thermal properties (σ/κ), and (d) ZT values of various oxygen-deficient $SrTiO_3$ samples [60].

(2.45×10^{-8} V^2K^{-2}), the influence of electronic (κ_e) and phononic (κ_L) to total thermal conductivity was calculated and observed that the lattice thermal conductivity is dominant as presented in **Figure 12(b)**. The lattice thermal conductivity was also plotted as a function of T^{-1} as shown in inset **Figure 12(b)**. The linear connection proposes that the lattice thermal conductivity is affected dominantly by Umklapp scattering [67]. Also, one should note that the similar grain sizes as shown in **Figure 8**, suggests that the reduction in thermal conductivity is mainly due to oxygen vacancy rather than grain boundary scattering.

Figure 12(c) illustrates the temperature-dependent PF for various oxygen-deficient $SrTiO_3$ samples. The substantial enhancement in the PF is due to the combination of moderate Seebeck coefficient values and high electrical conductivity. The PF obtained in this work is comparable to the PF obtained in doped-$SrTiO_3$ [65]. The DFT calculated PF for various carrier concentrations (10^{18}–10^{22}/cm^3) suggests that the highly reduced $SrTiO_3$ samples would have higher PF as shown in **Figure 12(d)** [68]. Additionally, one of the most remarkable features of this study is the decoupling among electrical and thermal conductivity of $SrTiO_{3-\delta}$ through solely oxygen vacancy as shown in **Figure 12(e)**. This opposite behavior led to a significant increase in the ratio of electrical to thermal conductivity (σ/κ), which is a key motive for TE materials. The complete outcome of oxygen vacancies in

Materials	Type	Maximum zT value	Working temperature (K)	References
Cd-doped $Ca_3Co_4O_9$	*p*	0.35	1000	[69]
Na_xCoO_2	*p*	0.31	800	[70]
Na_xCoO_2/Ag/Au	*p*	0.4–0.5	1000	[71]
$Li_{1-x}NbO_2$	*p*	0.125	970	This work
Nb-doped $SrTiO_3$	*n*	0.165	900	[72]
La-doped $SrTiO_3$ single crystal	*n*	0.27	1073	[73]
Pr-doped $SrTiO_3$	*n*	0.35	773	[74]
$SrTiO_{3-\delta}$	*n*	4.7×10^{-2}	670	This work

Table 2.
The dimensionless figure of merit zT of p-*type and* n-*type oxide thermoelectric materials for comparison.*

the TE materials can be represented by the temperature dependence of the ZT as shown in **Figure 12(f)**. It can be observed that the introduction of oxygen vacancy substantially increases the ZT value of 4.7×10^{-2} for [20HAr] reduced samples at 670 K. Therefore, we suggest that carefully adjusting the oxygen vacancy could be an effective strategy for oxide TE materials.

Table 2 shows that the efforts in the layer-structured cobaltites and $SrTiO_3$ based materials. It shows that significant improvements have been achieved in oxide TE materials, which could be of great interest for power generation applications at high operating temperatures. As compared with other oxides, the materials investigated in this study show relatively low figure of merit (zT), which is because the investigated $Li_{1-x}NbO_2$ and $SrTiO_{3-\delta}$ are pure, undoped materials to understand the mechanism for cation and anion defects effects.

3.3 Conclusion

In conclusion, this work demonstrated that cation and anion vacancies can successfully control the thermoelectric performance of oxide-based thermoelectric materials. These findings suggest that both cation defect and anion defect can be engineered by reducing atmospheres and the defects in oxide thermoelectric materials simultaneously act as a source of charge carriers and phonon scattering centers. This decoupled behavior between electrical conductivity and thermal conductivity can lead to a substantial increase in the thermoelectric performance of oxide materials. The concept applied in this work is generally important and has the possibility of impacting the thermoelectric performance of oxide thermoelectrics and other functional oxide materials.

Acknowledgements

The experimental work was conducted by using the facilities in the Korea Institute of Ceramic Engineering and Technology (KICET). The authors, Jamil Ur Rahman and Soonil Lee are grateful to all colleagues for help and support in the same research group at KICET.

Conflicts of interest

There are no conflicts of interest to declare.

Author details

Jamil Ur Rahman[1], Gul Rahman[2] and Soonil Lee[3]*

1 Department of Physics, Centre for Materials Science and Nanotechnology, University of Oslo, Sem Saelandsvei, Oslo, Norway

2 Department of Physics, Quaid-i-Azam University, Islamabad, Pakistan

3 School of Materials Science and Engineering/Department of Materials Convergence and System Engineering, Changwon National University, Changwon, Korea

*Address all correspondence to: leesoonil@changwon.ac.kr

References

[1] G. Mahan, B. Sales, J. Sharp, Thermoelectric materials: New approaches to an old problem. *Physics Today* **50**, 42-47 (1997).

[2] M. S. Dresselhaus *et al.*, New Directions for Low-Dimensional Thermoelectric Materials. *Adv. Mater.* **19**, 1043-1053 (2007).

[3] L.-D. Zhao, V. P. Dravid, M. G. Kanatzidis, The panoscopic approach to high performance thermoelectrics. *Energy Environ. Sci.* **7**, 251-268 (2014).

[4] T. M. Tritt, M. Subramanian, Thermoelectric materials, phenomena, and applications: a bird's eye view. *MRS bull.* **31**, 188-198 (2006).

[5] G. J. Snyder, E. S. Toberer, Complex thermoelectric materials. *Nat. Mater.* **7**, 105-114 (2008).

[6] A. Ioffe, *Semiconductor Thermoelements and Thermoelectric Cooling (London: Infosearch, 1957)*. Google Scholar, pp. 36-38.

[7] Z. Dughaish, Lead telluride as a thermoelectric material for thermoelectric power generation. *Physica B: Condens. Matter* **322**, 205-223 (2002).

[8] G. Nolas, J. Cohn, G. Slack, S. Schujman, Semiconducting Ge clathrates: Promising candidates for thermoelectric applications. *Appl. Phys. Lett.* **73**, 178-180 (1998).

[9] J. M. Song *et al.*, Chemically synthesized Cu2Te incorporated Bi-Sb-Te p-type thermoelectric materials for low temperature energy harvesting. *Scr. Mater.* **165**, 78-83 (2019).

[10] J. S. Yoon *et al.*, High thermoelectric performance of melt-spun $Cu_xBi_{0.5}Sb_{1.5}Te_3$ by synergetic effect of carrier tuning and phonon engineering. *Acta Mater.* **158**, 8 (2018).

[11] N. Van Du *et al.*, X-site aliovalent substitution decoupled charge and phonon transports in XYZ half-Heusler thermoelectrics. *Acta Mater.* **166**, 650-657 (2019).

[12] N. Van Du *et al.*, Synthesis and thermoelectric properties of Ti-substituted (Hf0. 5Zr0. 5) 1-xTixNiSn0. 998Sb0. 002 Half-Heusler compounds. *J. Alloys Compd.* **773**, 1141-1145 (2019).

[13] B. Sales, D. Mandrus, R. K. Williams, Filled skutterudite antimonides: a new class of thermoelectric materials. *Science* **272**, 1325-1328 (1996).

[14] H. Ohta, Thermoelectrics based on strontium titanate. *Mater. Today* **10**, 44-49 (2007).

[15] S. Lee, R. H. T. Wilke, S. Trolier-McKinstry, S. Zhang, C. A. Randall, $Sr_xBa_{1-x}Nb_2O_{6-\delta}$ Ferroelectric-thermoelectrics: Crystal anisotropy, conduction mechanism, and power factor. *Appl. Phys. Lett.* **96**, 031910 (2010).

[16] I. Terasaki, Y. Sasago, K. Uchinokura, Large thermoelectric power in $NaCo_2O_4$ single crystals. *Phys. Rev. B* **56**, R12685 (1997).

[17] T. Kawata, Y. Iguchi, T. Itoh, K. Takahata, I. Terasaki, Na-site substitution effects on the thermoelectric properties of $NaCo_2O_4$. *Phys. Rev. B* **60**, 10584 (1999).

[18] K. Fujita, T. Mochida, K. Nakamura, High-Temperature Thermoelectric Properties of $Na_xCoO_{2-\delta}$ Single Crystals. *Jpn. J. Appl. Phys.*, **40**, 4644-4647 (2001).

[19] S. Ohta, T. Nomura, H. Ohta, K. Koumoto, High-temperature carrier transport and thermoelectric properties

of heavily La- or Nb-doped doped $SrTiO_3$ single crystals. *J. Appl. Phys.* **97**, 34106-034110 (2005).

[20] Y. Wang, N. S. Rogado, R. J. Cava, N. P. Ong, Spin entropy as the likely source of enhanced thermopower in $Na_xCo_2O_4$. *Nature* **423**, 422-425 (2003).

[21] K. Takada *et al.*, Superconductivity in two-dimensional CoO_2 layers. *Nature* **422**, 53-55 (2003).

[22] M. Roger *et al.*, Patterning of sodium ions and the control of electrons in sodium cobaltate. *Nature* **445**, 631-634 (2007).

[23] H. Nakatsugawa, K. Nagasawa, Evidence for the two-dimensional hybridization in $Na_{0.79}CoO_2$ and $Na_{0.84}CoO_2$. *J. Solid State Chem.* **177**, 1137-1145 (2004).

[24] K. Takahata, Y. Iguchi, D. Tanaka, T. Itoh, I. Terasaki, Low thermal conductivity of the layered oxide (N a,Ca)Co_2O_4: Another example of a phonon glass and an electron crystal. *Phys. Rev. B* **61**, 12551 (2000).

[25] E. G. Moshopoulou, a. J. J. C. P. Bordet, Superstructure and superconductivity in $Li_{1-x}NbO_2$ ($x \approx 0.7$) single crystals. *Phys. Rev. B* **59**, 9590-9599 (1999).

[26] N. Kumada, S. Watauchi, I. Tanaka, N. Kinomura, Superconductivity of hydrogen inserted $LiNbO_2$. *Mater. Res. Bull.* **35**, 1743-1746 (2000).

[27] E. R. Ylvisaker, W. E. Pickett, First-principles study of the electronic and vibrational properties ofLiNbO2. *Phys. Rev. B* **74**, 075104 (2006).

[28] M. J. Geselbracht, T. J. Richardson, A. M. Stacy, Superconductivity in the layered compound Li_xNbO_2. *Nature* **345**, 324-326 (1990).

[29] N. Kumada *et al.*, Topochemical reactions of Li_xNbO_2. *J. Solid State Chem.* **73**, 33-39 (1988).

[30] J. D. Greenlee, W. L. Calley, W. Henderson, W. A. Doolittle, Halide based MBE of crystalline metals and oxides. *Phys. Status Solidi C* **9**, 155-160 (2012).

[31] J. U. Rahman *et al.*, The Synthesis and Thermoelectric Properties of *p*-Type $Li_{1-x}NbO_2$-Based Compounds. *J. Electron. Mater.* **46**, 1740-1746 (2017).

[32] T. Okuda, K. Nakanishi, S. Miyasaka, Y. Tokura, Large thermoelectric response of metallic perovskites: $Sr_{1-x}La_xTiO_3$ ($0{\sim}x{\sim}0.1$). *Phys. Rev. B* **63**, (2001).

[33] J. U. Rahman *et al.*, Grain Boundary Interfaces Controlled by Reduced Graphene Oxide in Nonstoichiometric SrTiO3-δ Thermoelectrics. *Scientific Reports* **9**, 8624 (2019).

[34] K. Van Benthem, C. Elsässer, R. French, Bulk electronic structure of $SrTiO_3$: Experiment and theory. *J. Appl. Phys.* **90**, 6156-6164 (2001).

[35] H. Ohta, K. Sugiura, K. Koumoto, Recent progress in oxide thermoelectric materials: *p*-type $Ca_3Co_4O_9$ and *n*-type $SrTiO_3$. *Inorg. Chem.* **47**, 8429-8436 (2008).

[36] H. Muta, K. Kurosaki, S. Yamanaka, Thermoelectric properties of reduced and La-doped single-crystalline $SrTiO_3$. *J. Alloys Compd.* **392**, 306-309 (2005).

[37] Y. Cui *et al.*, Thermoelectric Properties of Heavily Doped *n*-Type $SrTiO_3$ Bulk Materials. *J. Electron. Mater.* **38**, 1002-1007 (2009).

[38] S. R. Sarath Kumar, A. I. Abutaha, M. N. Hedhili, H. N. Alshareef, Effect of oxygen vacancy distribution on the thermoelectric properties of La-doped $SrTiO_3$ epitaxial thin films. *J. Appl. Phys.* **112**, 114104 (2012).

[39] S. Ohta, H. Ohta, K. Koumoto, Grain size dependence of thermoelectric performance of Nb-doped $SrTiO_3$

polycrystals. *J. Ceram. Soc. Jpn.* **114**, 102-105 (2006).

[40] Y. Lin *et al.*, Thermoelectric power generation from lanthanum strontium titanium oxide at room temperature through the addition of graphene. *ACS Applied Mater. Interfaces* 7, 15898-15908 (2015).

[41] Y. Wang *et al.*, Interfacial thermal resistance and thermal conductivity in nanograined $SrTiO_3$. *Appl. Phys Express* **3**, 031101 (2010).

[42] S. S. Kumar, A. Z. Barasheed, H. N. Alshareef, High temperature thermoelectric properties of strontium titanate thin films with oxygen vacancy and niobium doping. *ACS Applied Mater. Interfaces* **5**, 7268-7273 (2013).

[43] J. U. Rahman *et al.*, Localized double phonon scattering and DOS induced thermoelectric enhancement of degenerate nonstoichiometric $Li_{1-x}NbO_2$ compounds. *RSC Adv.,* 7, 53255-53264 (2017).

[44] P. Giannozzi *et al.*, QUANTUM ESPRESSO: a modular and open-source software project for quantum simulations of materials. *J. Phys. Condens. Matter* **21**, 395502 (2009).

[45] G. K. Madsen, D. J. Singh, BoltzTraP. A code for calculating band-structure dependent quantities. *Comput. Phys. Commun* **175**, 67-71 (2006).

[46] J. M. Ziman, *Principles of the Theory of Solids*. (Cambridge University Press, 1979).

[47] H. F. Roth, G. Meyer, Z. Hu, G. Kaindl, Synthesis, structure, and *X*-ray absorption spectra of Li_xNbO_2 and Na_xNbO_2 ($x \leq 1$). *Z Anorg. Allg. Chem.* **619**, 1369-1373 (1993).

[48] A. Miura *et al.*, Octahedral and trigonal-prismatic coordination preferences in Nb-, Mo-, Ta-, and W-based ABX_2 layered oxides, oxynitrides, and nitrides. *J. Solid State Chem.* **229**, 272-277 (2015).

[49] K. W. Lee, J. Kuneš, R. T. Scalettar, W. E. Pickett, Correlation effects in the triangular lattice single-band systemLi$_x$NbO$_2$. *Phys. Rev. B* **76**, 144513-144511 (2007).

[50] B. G. Streetman, S. Banerjee, *Solid State Electronic Devices*. (Prentice Hall New Jersey, 1980).

[51] B.-L. Huang, M. Kaviany, Ab initio and molecular dynamics predictions for electron and phonon transport in bismuth telluride. *Phys. Rev. B* 77, 125209 (2008).

[52] H. Kitagawa, A. Kurata, H. Araki, S. Morito, E. Tanabe, Structure and carrier transport properties of hot-press deformed Bi0. 5Sb1. 5Te3. *Phys. Status Solidi (a)* **207**, 401-406 (2010).

[53] T. Zhu *et al.*, Hot deformation induced bulk nanostructuring of unidirectionally grown p-type (Bi, Sb) 2 Te 3 thermoelectric materials. *J. Mater. Chem. A* **1**, 11589-11594 (2013).

[54] C. Kittel, *Introd. Solid State Phys.*, (Wiley, 2005).

[55] N. H. a. M. E. Fine, The Lorenz ratio and electron transport properties of NbO. *J. Appl. Phys.* **52**, 2876 (1981).

[56] W. Kohn, L. Sham, doi: 10.1103/PhysRev. 140. A1133. *Phys. Rev. A* **140**, 113 (1965).

[57] T. Mizokawa, T. Mizokawa and A. Fujimori. *Phys. Rev. B* **51**, 12 (1995).

[58] S. Okamoto, A. J. Millis, N. A. Spaldin, Lattice relaxation in oxide heterostructures: LaTiO 3/SrTiO 3 superlattices. *Phys. Rev. Lett.* **97**, 056802 (2006).

[59] H. Muta, K. Kurosaki, S. Yamanaka, Thermoelectric properties of rare earth doped $SrTiO_3$. *J. Alloys Compd.* **350**, 292-295 (2003).

[60] J. U. Rahman *et al.*, Oxygen vacancy revived phonon-glass electron-crystal in $SrTiO_3$. *J. Eur. Ceram. Soc.*, **39**, 8 (2019).

[61] C. Xu *et al.*, Formation mechanism of Ruddlesden-Popper-type antiphase boundaries during the kinetically limited growth of Sr rich $SrTiO_3$ thin films. *Scientific reports* **6**, 38296 (2016).

[62] F. Kröger, H. Vink, Relations between the concentrations of imperfections in crystalline solids. *Solid State Phys.* **3**, 307-435 (1956).

[63] S. Lee, G. Yang, R. H. T. Wilke, S. Trolier-McKinstry, C. A. Randall, Thermopower in highly reduced n-type ferroelectric and related perovskite oxides and the role of heterogeneous nonstoichiometry. *Phys. Rev. B* **79**, 134110-134111 (2009).

[64] P. Cox, *Transition metal oxides*. (Oxford University Press, 1992), pp. 158.

[65] A. V. Kovalevsky, A. A. Yaremchenko, S. Populoh, A. Weidenkaff, J. R. Frade, Effect of A-Site Cation Deficiency on the Thermoelectric Performance of Donor-Substituted Strontium Titanate. *J. Phys. Chem. C* **118**, 4596-4606 (2014).

[66] S. Piskunov, E. Heifets, R. Eglitis, G. Borstel, Bulk properties and electronic structure of SrTiO3, BaTiO3, PbTiO3 perovskites: an ab initio HF/DFT study. *Computational Materials Science* **29**, 165-178 (2004).

[67] J. M. Ziman, *Electrons and phonons: the theory of transport phenomena in solids*. (Oxford university press, 1960).

[68] J. Sun, D. J. Singh, Thermoelectric properties of *n*-type $SrTiO_3$. *APL Materials* **4**, 104803 (2016).

[69] S. Butt *et al.*, Enhancement of thermoelectric performance in Cd-doped $Ca_3Co_4O_9$ via spin entropy, defect chemistry and phonon scattering. *J. Mater. Chem. A* **2**, 19479-19487 (2014).

[70] K. Fujita, T. Mochida, K. Nakamura, High-temperature thermoelectric properties of $Na_xCoO_{2-\delta}$ single crystals. *Jpn. J. Appl. Phys.*, **40**, 4644 (2001).

[71] M. Ito, D. Furumoto, Effects of noble metal addition on microstructure and thermoelectric properties of $Na_xCo_2O_4$. *J. Alloys Compd.* **450**, 494-498 (2008).

[72] N. Wang, L. Han, H. He, Y. Ba, K. Koumoto, Effects of mesoporous silica addition on thermoelectric properties of Nb-doped $SrTiO_3$. *J. Alloys Compd.* **497**, 308-311 (2010).

[73] S. Ohta, T. Nomura, H. Ohta, K. Koumoto, High-temperature carrier transport and thermoelectric properties of heavily La-or Nb-doped $SrTiO_3$ single crystals. *J. Appl. Phys.* **97**, 034106 (2005).

[74] A. M. Dehkordi, S. Bhattacharya, J. He, H. N. Alshareef, T. M. Tritt, Significant enhancement in thermoelectric properties of polycrystalline Pr-doped $SrTiO_{3-\delta}$ ceramics originating from nonuniform distribution of Pr dopants. *Appl. Phys. Lett.* **104**, 193902 (2014).

Chapter 6

Thermoelectricity Properties of Tl_{10-x} ATe_6 (A = Pb) in Chalcogenide System

Waqas Muhammad Khan and Wiqar Hussain Shah

Abstract

The different elements are doping in the tellurium telluride to determine the different properties like electrical and thermal properties of nanoparticles. The chalcogenide nanoparticles can be characteristics by the doping of the different metals which are like the holes. We present the effects of Pb doping on the electrical and thermoelectric properties of Tellurium Telluride $Tl_{10-x}Pb_xTe_6$ (x = 1.000, 1.250, 1.500, 1.750, 2.000) respectively, which were prepared by solid state reactions in an evacuated sealed silica tubes. Structurally, all these compounds were found to be phase pure as confirmed by the x-rays diffractometery (XRD) and energy dispersive X-ray spectroscopy (EDS) analysis. The thermo-power or Seebeck co-efficient (S) was measured for all these compounds which show that S increases with increasing temperature from 295 to 550 K. The Seebeck coefficient is positive for the whole temperature range, showing p-type semiconductor characteristics. Similarly, the electrical conductivity (σ) and the power factors have also complex behavior with *Pb and Sn* concentrations. The power factor ($PF=S^2\sigma$) observed for $Tl_{10-x}Pb_xTe_6$ compounds are increases with increase in the whole temperature range (290 K-550 K) studied here. Telluride's are narrow band-gap semiconductors, with all elements in common oxidation states, according to (Tl^+) 9 $(Pb^{3+})(Te^{2-})$6. Phases range were investigated and determined with different concentration of *Pb and Sn* with consequents effects on electrical and thermal properties.

Keywords: Pb doping, Seebeck coefficient, electrical conductivity, power factor

1. Introduction

The thermo-electro-materials are now used as the renewable energy. It is used as the place of the coal, water tides, solar cells etc. The thermo electro-materials have more efficiency and reliable. Thermoelectric is one of the most important approaches in the solid state physics which can be converted the heat energy in the electrical energy, help to increase the efficiency, effectiveness and competency. Its importance is increase since last twenty years when the ease of use of fossil fuel is decrease. So there are different thermoelectric materials are used for the different temperatures from 10 K to the 1000 K which are used in the different applications for the cooling and heating [1–5]. Tellurium telluride is one important compound of the thermoelectric material which is studied, modified and increases the efficiency for the more and more applications for generation of power [1] and solar cells [2].

It is statistical results show that up to 60% of energy is losing in vain worldwide, most in the form of waste heat. High value of performance which is thermoelectric (TE) materials that have directly and inversely changed heat energy to electrical energy has thus drawn growing attentions of governments and research institutes [6]. Thermoelectric system is an environment-friendly energy conversion technology with the advantages of small size, high reliability, no pollutants and feasibility in a wide temperature range. However, the efficiency of thermoelectric devices is not high enough to rival the Carnot efficiency [7, 8]. Tellurium telluride is a basically alloy that is used for the increases the energy conversion efficiency at the any temperature of the heating and cooling in the electrical circuit [9, 10].

Many new thermoelectric materials or new material with which have high performance have been found such as skutterudites with high scattering rates of phonons [11, 12], silicon nanowires [13, 14], TE thin films [15], and nanostructured bismuth antimony telluride bulk alloys [16]. It was thinking that high barriers and extremely degenerately doped superlattices must achieve significant increases in thermos-electric power factor over bulk materials [17, 18]. It was revised that electron transport which are perpendicular to the barrier and investigated that large number of degenerate doped semiconductor or metal super-lattices could achieve which shows the power factors higher than the bulk and determined that non-maintenance of transverse momentum can have a large effect (especially in the case of metal super-lattices) by increasing the number of electrons contributing to conduction by thermionic emission [19].

An electric field provides a potential difference along a wire of electrons, which creates a force,

$$\mathrm{F} = eE \tag{1}$$

where 'e' is the charge of an electron and 'E' is the magnitude of the electric field. That force accelerates the electrons, as expected by Newton's second law. We get the expression for v_d by equating the above two equations of force,

$$\mathrm{F} = \mathrm{m_e a} \tag{2}$$

We get the expression for v_d by equating the above two equations of force,

$$\mathrm{m_e a} = eE \tag{3}$$

As we know that

$$\mathrm{a} = \frac{vd}{\tau} \tag{4}$$

$$\frac{mevd}{\tau} = eE \tag{5}$$

$$v_d = \frac{Ee\tau}{m_e} \tag{6}$$

Plugging this expression into that of conductivity, we get

$$\sigma = \frac{nev_d\mathrm{A}}{AE} \tag{7}$$

$$\sigma = \frac{nev_d}{E} \tag{8}$$

Using the equation of drift velocity.

$$v_d = \frac{Ee\tau}{m_e} \tag{9}$$

We have

$$\sigma = \frac{ne}{E}\frac{Ee\tau}{m_e} \tag{10}$$

$$\sigma = \frac{\text{ne}^2\tau}{m_e} \tag{11}$$

This is the required expression of electrical conductivity.
The figure of merit is

$$ZT = \frac{S^2\sigma T}{k} \tag{12}$$

Where is σ is the electrical conductivity, *k* is the thermal conductivity, S is the see-beck coefficient, and T is the absolute temperature which is determined the efficiency of the thermo electric materials applications [8]. The power factors can be determined the electrical and thermal properties. The power factor can be defined as $S^2\sigma$. It can be help the determination of the charge carrier's concentration, from the doping concentration charges and lay down the free electrons in the system of chalcogenides.

We have investigated the chalcogenide with different materials (lead, tin, bismuth etc.) doped in the thallium tellurides. They have complex composition and structure on the basis of the electronic configurations. These compositions help to increases their properties like thermal, electrical, optical etc. of the thermo-electrical materials. There are many challenges of complex composition to high their electrical conductivity, high see-beck coefficient and low thermal conductivity. Due to this, they can controlled the electronic structures of the system i.e. band gap, shapes and degenerated level which is near the Fermi level, concentration of electrons and charge carriers scattered depend on them [7, 8].

The ideal situation is having high effective mass and high mobility, but this is extremely difficult to tune in a material. One has to compromise one for the other, and there are reports of compounds showing promising thermoelectric material at both ends of the spectrum. Chalcogenides (group 16 10 elements).

2. Experimental section

The Pb doped $Tl_{10-x}A_xTe_6$ (A = Pb) is (x = 1.00, 1.25, 1.50, 1.75, 2.00) has been prepared by solid state reactions in evacuated sealed silica tubes. The purpose of this study were mainly for discovering new type of ternary compounds by using Tl^{+1}, $Pb^{+3}$and Te^{-2} elements as the starting materials. Direct synthesis of stoichiometric amount of high purity elements i.e. 99.99% of different compositions have been prepared for a preliminary investigation. Since most of these starting materials for solid state reactions are sensitive to oxygen and moistures, they were weighing stoichiometric reactants and transferring to the silica tubes in the glove box which is filled with Argon. Then, all constituents were sealed in a quartz tube.

Before putting these samples in the resistance furnace for the heating, the silica tubes was put in vacuum line to evacuate the argon and then sealed it. This sealed power were heated up to 650 C° at a rate not exceeding 1 k/mint and kept at that temperature for 24 hours. The sample was cooled down with extremely slow rate to avoid quenching, dislocations, and crystals deformation.

Structural analysis of all these samples was carried out by x-rays diffraction, using an Inel powder diffractometer with position-sensitive detector and CuKα radiation at room temperature. No additional peaks were detected in any of the sample discussed here. X-ray powder diffraction patterns confirm the single phase composition of the compounds.

The temperature dependence of Seebeck co-efficient was measured for all these compounds on a cold pressed pellet in rectangular shape, of approximately $5 \times 1 \times 1\ mm^3$ dimensions. The air sensitivity of these samples was checked (for one sample) by measuring the thermoelectric power and confirmed that these samples are not sensitive to air. This sample exposes to air more than a week, but no appreciable changes observed in the Seebeck values. The pellet for these measurements was annealed at 400°C for 6 hours.

For the electrical transport measurements 4-probe resistivity technique was used and the pellets were cut into rectangular shape with approximate dimension of $5 \times 1 \times 1\ mm^3$.

3. Result and discussions

3.1 Structural analysis

X-ray diffraction is used for the structural analysis of the materials. It helps to determine the crystal structure and particle size. Several of $Tl_{10-x}ATe_6$ which have doped Pb and Sn in it. The A is the different doped element. Here, A is Pb and Sn. It has the different concentration of it. The X-ray diffraction of $Tl_{10-x}ATe_6$ with different concentration of doping of Pb and Sn is shown in **Figure 1**. Due to the different concentration, their peaks are different shown in **Figure 1**. Crystal size which prepared samples in the range of (20-24 nm) indicating that Pb & Sn incorporations in $TlATe_2$ do not affect significantly its crystallite size.

The average particle size (D) was estimated by Scherrer formula given as

$$D = \frac{k\lambda}{\beta cos\theta} \tag{13}$$

where 'k' is the Scherrer's constant and also known as shape constant having value 0.9 in our case, 'λ' is wavelength of x-rays, 'β' is full width at half maximum (FWHM) and θ is the Bragg angle.

By using the following equation, lattice parameters 'a', b and 'c' of cubic $Tl_{10-x}ATe_6$ were calculated.

$$\frac{1}{d^2} = \frac{h^2 + k^2 + l^2}{a^2} \tag{14}$$

For cubic structure, the volume can be calculated by

$$V = a^3 \tag{15}$$

and X-ray density can be calculated by

$$\rho = \frac{nM}{VN_A} \tag{16}$$

M is the molecular mass of $Tl_{10-x}ATe_6$, N_A is Avogadro's number and its value is constant (6.02×10^{23} mol^{-1}). n is the number of atoms per unit cell.

The XRD data are summarized in **Table 1**.

Figure 2 shows the EDX of the $Tl_{10-x}XTe_6$, have the different concentration of the doping of the Pb in it. The EDX shows the composition of the compounds. It shows the Pb are present in it.

3.2 Electrical and thermal properties

To determine the different concentration of the doping of the Pb in the compound, there is changing in the charges carries. So the doping is effect on the temperature. Due to this temperature, it is variant in the Seebeck coefficient (S) as shown in **Figure 3**. The Seebeck coefficient can determined the temperature gradient for 1 K. It shows that the positive Seebeck effect from the 300 K to 500 K, for all p type semiconductors whose have the high charge carrier concentration. The Seebeck is positive due the concentration of doping elements is increase. So the mostly thermoelectric materials are the p type semiconductors materials. Due to increasing the concentration of doping elements, It improves the (i) reducing of grain size (ii) charge mobility and carrier density in thermos electric materials.

The Seebeck coefficient is varying from 80 to 120 μV/K as a function of the temperature. The behavior of the Seebeck coefficient is increasing as the Fermi level energy is decreasing due to the charge carrying density. In **Figure 4** shows that there is low level of charge carrier so that the holes are increase in it, so that it shows the high value of thermopower. So the large value of X, the doping elements have the large number of electrons and less number of charges carriers. As electrical conductivity 'σ' increases with increases in *n* according to the equation

$$\sigma = n e \mu \tag{17}$$

where
μ = carrier mobility
and
e = charge of carriers

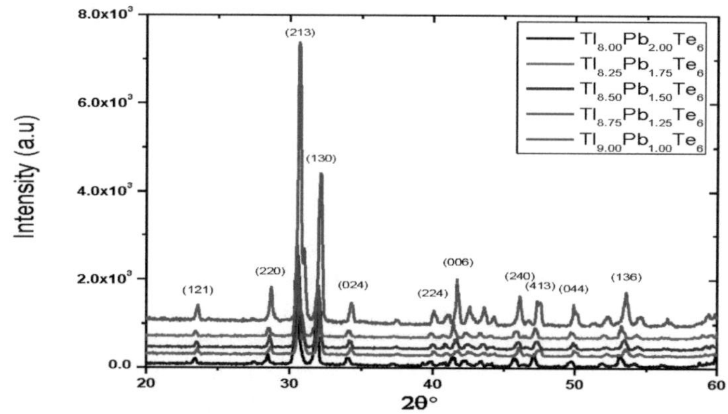

Figure 1.
XRD of doping of Pb in the TlATe (A = Pb).

Sample	Crystallite size, D = 0.9λ/βcosθ (nm)	Lattice constant a, b, c = (Å)	Volume ($Å^3$)
$Tl_9Pb_1Te_6$	32.247	a = b = 8.8931 c = 13.0052	1004.521
$Tl_{8.75}Pb_{1.25}Te_6$	31.795	a = b = 8.84510 c = 13.07515	1023.925
$Tl_{8.50}Pb_{1.50}Te_6$	31.180	a = b = 8.82510 c = 13.00010	1013.429
$Tl_{8.25}Pb_{1.75}Te_6$	31.128	a = b = 8.81010 c = 13.0010	1009.093
$Tl_8Pb_2Te_6$	31.055	a = b = 8.84814 c = 13.16215	1022.722

Table 1.
Value of crystallite size and crystal system of doped $Tl_{10-x}A_xTe_6$ (A = Pb).

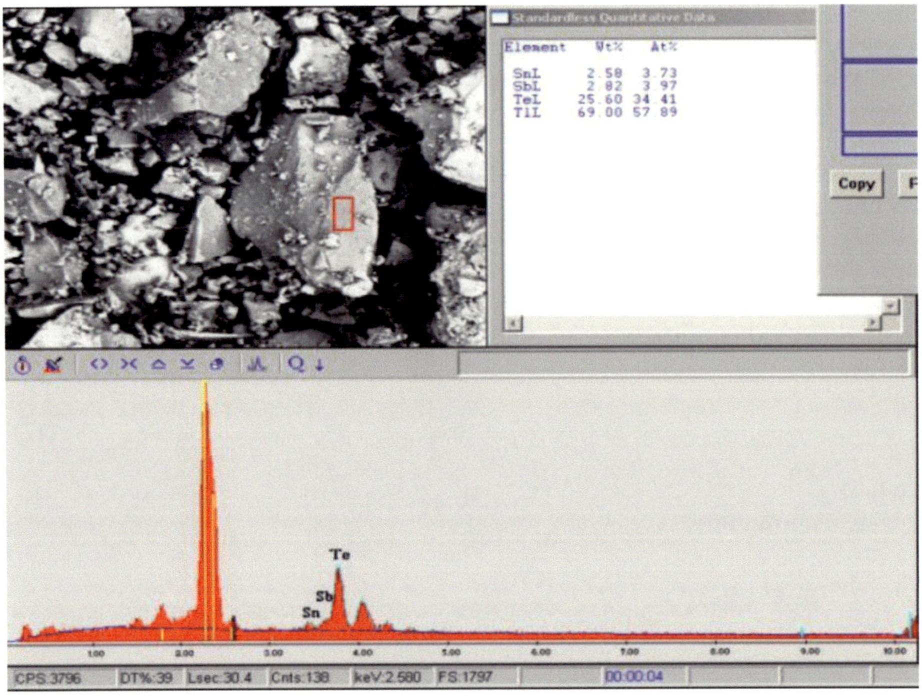

Figure 2.
Comparison of the EDX of doping of Pb in the TlATe (A = Pb).

In **Figure 4** shows, the electrical conductivity of the quaternary compounds as compared to the temperature while the temperature is varied. The electrical conductivity is decrease as the temperature is increase that is why it is show the p type semiconductor and behave the positive temperature coefficient. It is cause the phonons scattering the charge carriers and effects the grains boundary. As increase the doping of elements, the holes in the compounds are increase, which is cause the phonons scattering. In chalcogenide system, the different elements are doping in the compound has no effect on the electrical conductivity. The low electrical conductivity is due to the effect oxide as the impurity in the compounds.

The Electrical Conductivity data are summarized in **Table 2**.

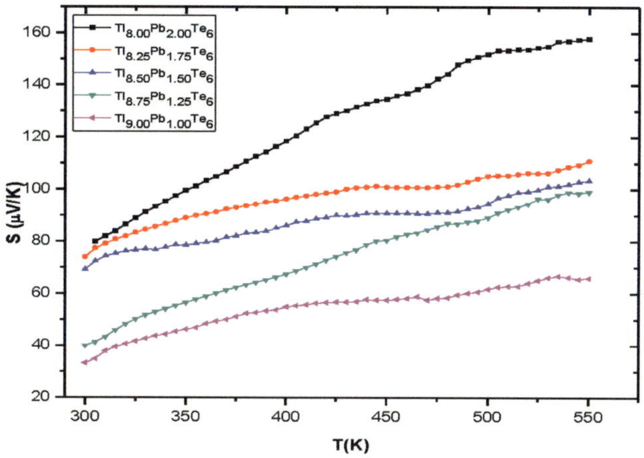

Figure 3.
The see-Beck Co-efficient of doping of Pb in the TlATe (A = Pb).

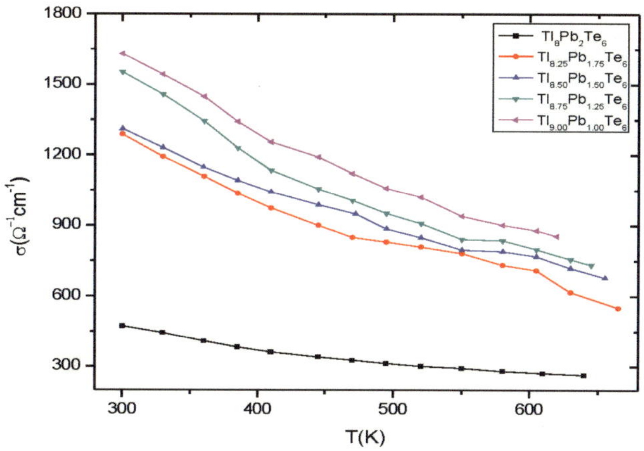

Figure 4.
The electrical conductivity of doping of Pb in the TlATe (X = Pb).

Sample	Electrical conductivity (Ω^{-1} cm^{-1}) at 300 K	Electrical conductivity (Ω^{-1} cm^{-1}) at 550 K
$Tl_9Pb_1Te_6$	1645	890
$Tl_{8.75}Pb_{1.25}Te_6$	1540	750
$Tl_{8.50}Pb_{1.50}Te_6$	1335	610
$Tl_{8.25}Pb_{1.75}Te_6$	1301	585
$Tl_8Pb_2Te_6$	460	288

Table 2.
The electrical conductivity of $Tl_{10-x}A_xTe_6$ (A = Pb) at 300 and 550 K for all samples ($1.0 \le x \le 2.0$).

The behavior of temperature is different for the different concentration of the compound. The relationship between the Seebeck, temperature and concentration of doping elements as given below.

$$S = \frac{8\pi^2 k_B^2}{3eh^2} m^* T \left(\frac{\pi}{3n}\right)^{\frac{2}{3}} \tag{18}$$

Where, k_B is the Boltzmann constant, e is the electronic charge, h is the Planck's constant, m^* is the effective mass and n is the charge carrier concentration. The effective mass and concentration are two parameters of the Seebeck coefficient. The samples have low concentration, it increase the thermos-power as well as the temperature.

Figure 4 shows that the electrical conductivity σ is decrease as the increase of the temperature of the compounds. Increased, doping concentration causes decrease in ó as expected and inversely affecting their Seebeck counterpart [20]. The Seebeck is inversely effect due to the increasing of the doping of the concentration of the doping.

The Seebeck coefficient data are summarized in **Table 3**.

The different compounds have enhance the power factor

$$PF = S^2\sigma \tag{19}$$

is decreases the electrical conductivity as increases the Seebeck coefficient in the given system. The PF is depend on the Seebeck coefficient. To measure the PF by the knowing the electrical conductivity and Seebeck coefficient in **Figure 5**. As increases the temperature, the power factor is increases for all the compounds.

Sample	See-beck coefficient(μVK^{-1}) at 300 K	See-beck coefficient(μVK^{-1}) at 550 K
$Tl_9Pb_1Te_6$	32	56
$Tl_{8.75}Pb_{1.25}Te_6$	40	90
$Tl_{8.50}Pb_{1.50}Te_6$	68	100
$Tl_{8.25}Pb_{1.75}Te_6$	73	110
$Tl_8Pb_2Te_6$	80	160

Table 3.
See-beck Co-efficient of all doped $Tl_{10-x}A_xTe_6$ (A = Pb) samples at 300 and 550 K.

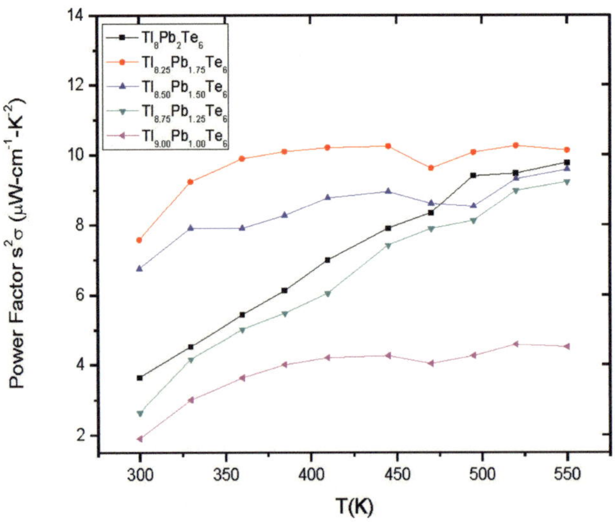

Figure 5.
The power factor of doping of Pb in the TlATe (A = Pb).

Sample	Power factor $S^2\sigma$ (μWcm^{-1} K^{-2}) at 300 K	Power factor $S^2\sigma$ (μWcm^{-1} K^{-2}) at 550 K
$Tl_9Pb_1Te_6$	1.8	4.7
$Tl_{8.75}Pb_{1.25}Te_6$	2.8	8.5
$Tl_{8.50}Pb_{1.50}Te_6$	6.8	10
$Tl_{8.25}Pb_{1.75}Te_6$	7.9	10.2
$Tl_8Pb_2Te_6$	3.6	9.7

Table 4.
Power factor of all doped $Tl_{10-x}A_xTe_6$ (A = Pb) samples at 300 and 550 K.

Figure 5 shows that the power factor increase as the doping of the concentration is increases. As increases, the doping of the elements in the compounds is increase the optimization, which can help to increases the Seebeck and power factor.

The power factor data are summarized in **Table 4**.

4. Conclusion

The different concentration of the doped Pb in the $Tl_{10-x}ATe_6$ nanoparticles is synthesized by the solid state reaction within evacuated silica sealed tube with the pellet size is 5*1*1 mm^3 in the rectangular dimension and then studied the electrical and thermal properties of the nanoparticles. The XRD shows the nanoparticles are the single phase, crystal structure measured by the experimental formula, having the same space group 14/mcm like Tl_5Te_3. The doping of the holes materials it changes its physical properties i.e. thermal, electrical, phase etc. For the electrical transport measurements 4-probe resistivity technique was used and the pellets were cut into rectangular shape with approximate dimension of 5 × 1 × 1 mm^3. Due the doping of Pb in the $Tl_{10-x}ATe_6$ nanoparticles the Seebeck coefficient is increases. The phase of the both nanoparticles is also change. The phase is come to the face centered cubic. It is also shows that the increases temperature decreases the electrical conductivity due to the doping of Pb in the $Tl_{10-x}ATe_6$ nanoparticles. The power factor is increases because the Seebeck coefficient is increase.

Author details

Waqas Muhammad Khan* and Wiqar Hussain Shah
Department of Physics, Faculty of Basic and Applied Science, International Islamic University, H-10, Islamabad, Pakistan

*Address all correspondence to: waqaskhanrwp@gmail.com

References

[1] Caillat T, Fleurial J, Borshchevsky A. AIP conf. Proc. 1998;**420**:1647

[2] Campana RJ. Adv. Ener. Conv. 1962;**2**:303

[3] Mehta RJ, Zhang Y. C. Karthika eta. Nature Materials. 2012;**11**:233-240

[4] G.S. Nolas, J. Poon and M. Kanatzidis, MRS, Bull **31**199 (2006),

[5] Kuropaatawa BA, Assoud A, Klienke H. J. Alloys and Compounds. 2011;**509**:6768

[6] Kanatzidis MG. Chemistry of Materials. 2010;**22**:648

[7] LaLonde AD, Pei YZ, Wang H, Snyder GJ. Materials Today. 2011;**14**:526

[8] Nolas GS, Morelli DT, Tritt TM. Annual Review of Materials Science. 1999;**29**:89

[9] K. Kurosaki, A. Kosuge, H. Muta, M. Uno, and S. Yamanaka, Applied Phys. Letts 87, 06191 (2005)

[10] Yang J, Stablers FR, Electrn J. Maternité. 2009;**38**:1245

[11] Synder GJ, Toberer ES. Nature Materials. 2008;**7**:105

[12] Soostsman JR, Chung DY, Kanatzdis MG. Angew Chem. Inter. Ed. **48**. In: 8616. 2009

[13] Yang J, Zhang W, Bai SQ, Mei Z, Chen LD. Applied Physics Letters. 2007;**90**:108

[14] R. Chen, A. I. Hochbaum, R. D. Delgado, W. Liang, E. C. Garnett, M. Najarian, A. Majumdar, P. Yang, Enhanced thermoelectric performance in rough silicon nanowires, in: APS Meet., 2008. <http://meetings.aps.org/Meeting/MAR08/Event/77251> (accessed December 4, 9(2017).

[15] Boukai AI, Bunimovich Y, Tahir-Kheli J, Yu J, Goddard WAI, Heath JR. Nature. 2008;**451**:168

[16] Venkatasubramanian R, Siivola E, Colpitts T, O'Quinn B. Nature. 2001;**413**:597

[17] Poudel B, Hao Q, Ma Y, Minnich A, Muto A, Lan YC, et al. Science. 2008;**320**:634

[18] Shakouri A, Labounty C, Abraham P, Piprek J, Bowers JE. Materials Research Society Proceedings. 1999;**545**:449

[19] Vashaee D, Shakouri A. Physical Review Letters. 2004;**92**:106103/1

[20] Wiqar H. Shah, Aqeel Khan, Waqas Khan, Waqar Adil Syed. Chalcogenide Letters. 2017;**14**:61

Chapter 7

Thermoelectric Elements with Negative Temperature Factor of Resistance

Yuri Bokhan

Abstract

The method of manufacturing of ceramic materials on the basis of ferrites of nickel and cobalt by synthesis and sintering in controllable regenerative atmosphere is presented. As the generator of regenerative atmosphere the method of conversion of carbonic gas is offered. Calculation of regenerative atmosphere for simultaneous sintering of ceramic ferrites of nickel and cobalt is carried out. It is offered, methods of the dilated nonequilibrium thermodynamics to view process of distribution of a charge and heat along a thermoelement branch. The model of a thermoelement taking into account various relaxation times of a charge and warmth is constructed.

Keywords: thermoelement, ferrites, thermistor, nanoceramics, model

1. Introduction

Perspective direction of building of thermal cells with a high Z-factor is use nanomaterial's and nanocomposite [1].

The quantity of works, theoretical and experimental, devoted to research thermoelectric nanomaterial's, steadily grows. The received results are optimistically enough, at least, from the point of view of fundamental science.

The augmentation of thermoelectric quality factor in nanomaterial's is bound to two physical phenomena [2]:

- The thermal conductivity decreases, bound to presence of numerous borders of partition. And it is necessary to notice that basically such reduction is bound with фононной thermal conductivity and make small impact on electronic transport;

- Augmentation of width of the forbidden area in ceramic наноматериалах with simultaneous augmentation of density of states about Fermi's level. In this case there is an electric conductivity reduction. However thermal conductivity reduction, under certain conditions, can compensate such reduction and lead to quality factor augmentation.

For building of ceramic semi-conductor stuffs, instead of doping by certain impurities it is possible to use a furnacing method in a reducing atmosphere [3, 4]. Such method, together with use of stuffs with NTRC, will allow to raise quality factor of thermoelectric stuffs and to raise efficacy of thermal cells, especially in the field of high temperatures where ceramic stuffs, are the steadiest. Such method probably building n - and p - phylum of conductivity a choice of composition of gas atmosphere and furnacing temperatures in one shot.

2. Material's

In thermoelectric nanocomposites the size of grain does not exceed several tens nanometers. It is obvious that for increasing of thermoelectric performance efficiency, the following condition is necessary: the size of grain should be less, than the average length of free run of phonons, but more than the average length of free run of charge carriers (electrons or holes). In this case phonons on intercrystallite borders, disperse more effectively and it leads to stronger reduction of thermal conductivity (at the expense of reduction of grid contribution), in comparison with electric conductivity reduction, providing total increase of thermoelectric quality factor.

It is obvious that in nanocomposites the lobe of intercrystallite borders will increase with the reduction of aggregate size, it will lead to consecutive depressing of thermal conductivity of the material. It is natural that dispersion of electrons on intercrystallite borders will take place leading to the reduction of their motility. However, the thermal conductivity reduction in volume nanocomposites can be more essential, than electric conductivity reduction. Thus, volume nanocomposites, consisting from nano the grains of the thermoelectric material that are divided by intercrystallite borders, can potentially possess high thermoelectric efficiency. They will have electric conductivity high and low thermal conductivity simultaneously.

Oxidic ceramic materials may be transferred into semiconductor state by means of process of operated valence. For this purpose, different methods are used, such as, a restoration method, i.e. ceramics furnacing in the regenerative medium [3, 4], unisovalent substitution are routinely used. In this case it is possible to receive comprehensible conductivity, at conservation of low thermal conductivity. In ceramic materials probably substantial growth of the dispersion mechanism, at low thermal conductivity, but it leads to quality factor augmentation.

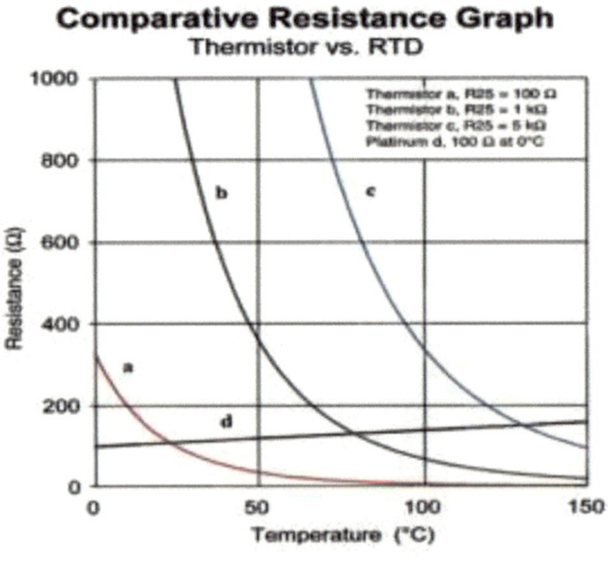

Figure 1.
Typical resistance dependence on the temperature for a thermistor with NTCR.

So, for example, some series of spinel's, at isomorphous substitution are transferred in to a state of the semi-conductor conductivity, possessing conductivity with NTRC (**Figure 1**) [5–8].

Oxide doping of nickel by lithium leads to sharp augmentation of conductivity at the expense of changing lithium ions into nickel ions in octahedral positions. The formation of the solid solution with uncompensated charge allows to create different types of conductivity by the variation of lithium concentration.

$$x/2\ Li_2O + (1-x)\ NiO + x/4\ O_2 \rightarrow \left(Ni^{2-}{}_{1-2x}\ Li^{+}{}_{x}\ Ni^{3+}{}_{x}\right) O$$

Similar reaction takes place when CO is replaced. If it is possible to omit superfluous oxygen, the solution of oxides receives the additional uncompensated charge in octahedrons spinel, and that process leads to conductivity augmentation. It follows that having combined doping with roasting in the recovery medium, it is possible to receive ceramic materials with adjustable conductivity. You should mind that thermal ceramic conductivity is defined by the phonon mechanism with characteristic wave length $\sim$ 5–10 microns, and creating necessary grain frame of ceramics it is possible to achieve substantial increase of quality factor of the material.

Joint roasting in the recovery medium of ferrite on the basis of spinel and dielectrics, such as solid solutions on the basis of perovskite families of titanates of barium, strontium and lead, allows to create multilayered frames [3].

Let us consider as model of spinels recovery $NiMn_2O_4$ and $CoMn_2O_4$ in a gaseous medium received by conversion CO_2 and H_2O over Carboneum. The calculation shows that such reactions happen in some stages. It is difficult to analysis all reactions, we will illustrate calculation on the reactions defining composition of a gaseous environment (**Figure 2**).

Reaction $CO_2 + C \rightarrow 2CO$ takes place with ΔZ_0=43462+6,121TlgT-62,746 T – is potential change of Gibbs at temperature T and $lgK_p = -9500\ T^{-1}$ - 1,338lgT + 13.715 - an equilibrium constant. Reaction $C + 2H_2O \rightarrow CO + H_2$ has with ΔZ_0=35730 +6,121TlgT-55,403 T and $lgK_p = -7811\ T^{-1}$ - 1,338lgT + 12.170 accordingly [6]. The joint analysis of the received expressions shows that thermodynamically reactions are resolved with $T \approx 500K$, and kineticly proceed with sufficient speed with $T \approx 700K$.

The recovery of nickel oxides and cobalt thermodynamically is possible with $T \approx 700K$. Thus, kineticly, recovery reactions take place in demanded atmosphere and there is no necessity of a pre-treatment of medium. Using diagrammes of Ellingem-Richardson-Dzheffez (**Figure 3**), we see that at the yielded temperatures cobalt and nickel recovery descends simultaneously. It allows to create simple adjustment of temperature necessary degree of restoration in spinels.

It is necessary to notice that at such temperatures there is a restoration to manganese metal. Calculation shows that manganese is reduced to the bivalent state and does not variate the position in a grid. At the same time nickel and cobalt, are restored to a monovalent state and provide n and p conductivity accordingly.

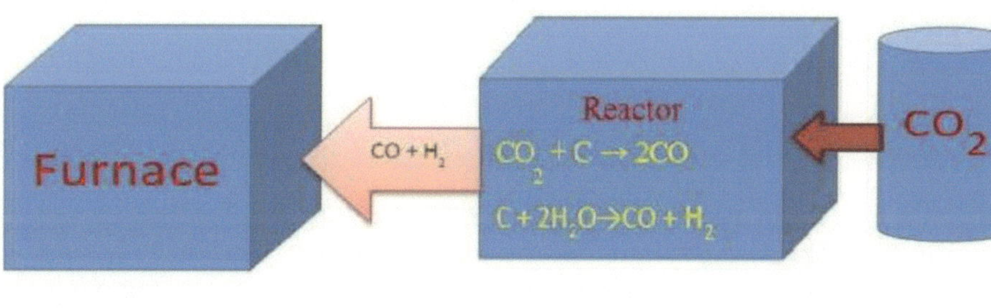

Figure 2.
The schema of installation for roasting in a reducing atmosphere.

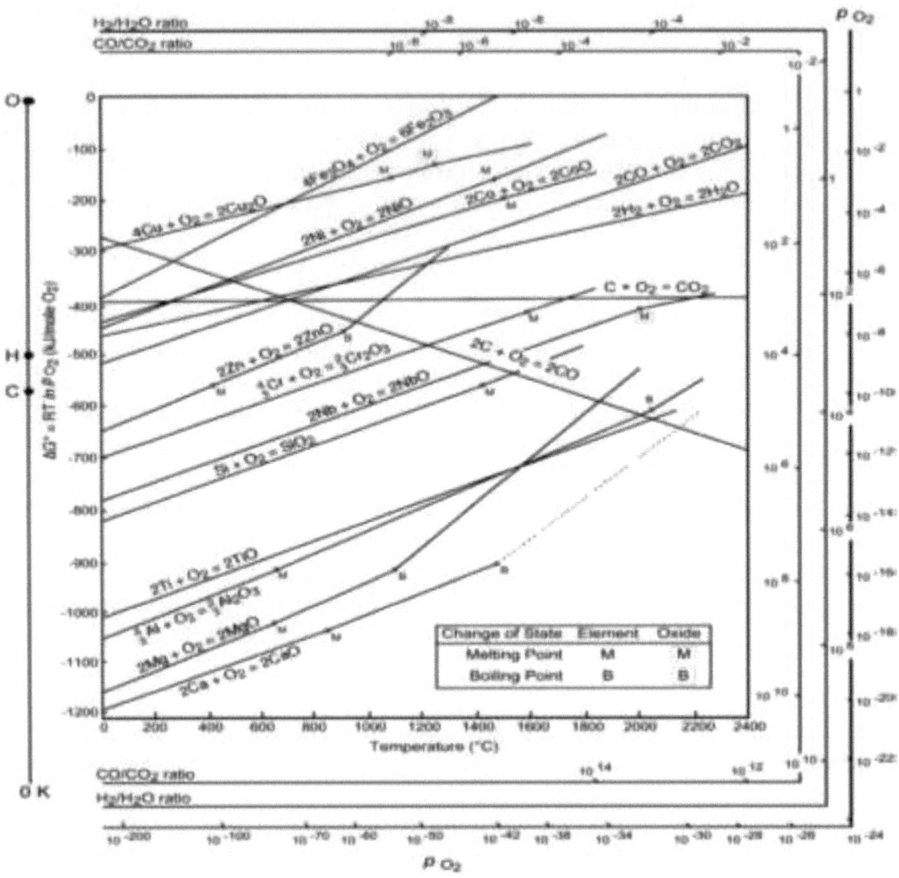

Figure 3.
Diagramme of Ellingem-Richardson-Dzheffez for equilibriums metal - metal oxide: Points of M and I_n - temperatures of elements phase changes.

Thus, it is possible to create semi-conductor branches of thermal cells with various types of conductivity in one technological process is possible.

It is obvious that when the size of grain will decrease, the area of borders of grains will increase. It will lead to higher degree of restoration of a ceramic material on border of grains. Accordingly, conductivity of borders will differ from conductivity actually grains. Naturally, dispersion electron's in grain borders will take place, and their mobility will be reduced. However, on the average, change of conductivity of a material will not be so essential because of certain uniformity of a material. Thus, there is a possibility at the expense of essential reduction phonon's heat conductivity to raise good quality of a material. Therefore, use of ceramic materials with NTCR allows to create high-temperature thermoelements with good quality [9].

3. Microscopic model

Known microscopic models of thermoelectric effect in semiconductors, to some extent, are grounded on the solution of the equation of Boltzmann's.

Let us spot thermoelectromotive force (TEF) in the presence of a temperature lapse rate ($\nabla T \neq 0$) from the stationary kinetic equation of Boltzmann's in relaxation time approach:

$$(\bar{V}, \nabla_{\bar{r}} f) + \frac{1}{\hbar}(\bar{F}, \nabla_{\bar{r}} f) = -\frac{f - f_0}{\tau(\bar{k})} = -\frac{f^{(1)}(\bar{k})}{\tau(\bar{k})}. \tag{1}$$

were $f - f_0 = f^{(1)}(\bar{k})$, a F – exterior force, $\tau(\bar{k})$ – a relaxation time.

In case of the undegenerated semiconductor the equilibrium distribution function for electrons is accepted in a view:

$f_{on} = e^{-\frac{E-\mu}{k_0 T}}$ for electrons,

$f_{op} = e^{\frac{\mu'-E}{k_0 T}} = e^{\frac{-E+\mu+\Delta E}{k_0 T}}$ for electron defects.

Here $\mu' = -\Delta E - \mu$, $E = \frac{{}^2 k'^2}{2m_p^*}$,.

Approving of the allowance to equilibrium function $f^{(1)}(\bar{k})$ by a small, on the left of Eq. (1) exchange f на f_0. Then we will have:

$$\nabla_{\bar{r}} f \approx \nabla_{\bar{r}} f_0 = \frac{\partial f_0}{\partial T}\nabla T + \frac{\partial f_0}{\partial \mu}\nabla \mu = \frac{\partial f_0}{\partial E}\left(\frac{\mu - \mathrm{E}}{T}\nabla T - \nabla \mu\right), \tag{2}$$

$$\nabla_{\bar{k}} f \approx \nabla_{\bar{k}} f_0 = \frac{\partial f_0}{\partial E}\nabla_{\bar{k}} \mathrm{E} = \hbar \frac{\partial f_0}{\partial E}\bar{\mathrm{V}}. \tag{3}$$

Substituting expressions (2) and (3) in the Eq. (1), and being restricted to a case when the electric field $\overline{E} = -\nabla \varphi$, where φ - electrostatic potential, we will have for electrons operates only:

$$\bar{V}_n \frac{\partial f_0}{\partial \mathrm{E}}\left\{\frac{\mu - \mathrm{E}}{T}\nabla T - \nabla \mu\right\} + e\frac{\partial f_0}{\partial \mathrm{E}}\bar{V}_n \nabla \phi = -\frac{f^{(1)}(\bar{k})}{\tau_e(\bar{k})}. \tag{4}$$

From the Eq. (4) we will spot $f_n^{(1)}(\bar{k})$:

$$f_n^{(1)}(\bar{k}) = \tau_e(\bar{k})\frac{\partial f_0}{\partial \mathrm{E}}\left\{\frac{\mu - \mathrm{E}}{T}\nabla T - \nabla(\mu - e\phi)\right\}\bar{V}_n. \tag{5}$$

for electron defects:

$$f_p^{(1)}(\bar{k}') = \tau_p \frac{\partial f_0}{\partial \mathrm{E}}\left\{\frac{\mathrm{E} + \mu + \Delta\mathrm{E}}{T}\nabla T - \nabla(\mu - e\phi)\right\}\bar{V}_p. \tag{6}$$

From relations (2)–(6) it is visible that the equilibrium distribution function is supposed nonuniform, i.e. in an equilibrium state there are processes of transport of heat and particles, and transport of particles is carried out not only at the expense of an external field. Usually it is considered that the dispersion mechanism can be considered through a relaxation time.

For example, dispersion of charge carriers is carried out at interaction with ultrasonic oscillations of a crystalline lattice. In this case the free length $l = V\tau$ does not depend on energy of carriers, and it is possible to express a relaxation time through l:

$$\tau = \frac{l}{V} = \frac{m_n^* l}{\hbar}k^{-1}. \tag{7}$$

Having entered a label $E = k_0 T\alpha$, here $\alpha = \frac{d\varepsilon}{dT}$ – specific TEF, equal to the relation TEF to an individual difference of temperature.

Then

$$\bar{j}_n = nu_n\left\{\nabla(\mu - e\phi) + \left(2k_0 - \frac{\mu}{T}\right)\nabla T\right\}, \tag{8}$$

$$\bar{j}_p = pu_p\left\{\nabla(\mu - e\phi) - \left(2k_0 - \frac{\mu + \Delta E}{T}\right)\nabla T\right\}. \tag{9}$$

In this expression $\frac{4el}{3\left(2\pi m_n^* k_0 T\right)^{1/2}} = u_n(T)$ – Mobility of electrons.

The full density of a current we will spot expression:

$$\bar{j} = nu_n\left\{\nabla(\mu - e\phi) + \left(2k_0 - \frac{\mu}{T}\right)\nabla T\right\} +$$

$$pu_p\left\{\nabla(\mu - e\phi) - \left(2k_0 - \frac{\mu + \Delta E}{T}\right)\nabla T\right\} \tag{10}$$

For a finding TEF it is necessary to spot a potential difference at a broken circuit. Having equated $\bar{j} = 0$, from (10) equality follows:

$$\nabla\left(\frac{\mu}{e} - \phi\right) = -\frac{k_0}{e}\left(2 - \frac{\mu}{k_0 T}\right)\nabla T.$$

Specific TEF α it is spotted as

$$\alpha = \frac{\left|\nabla\left(\phi - \frac{\mu}{e}\right)\right|}{|\nabla T|}. \tag{11}$$

For the natural semiconductor $n = p = n_i$, $\mu = -\frac{\Delta E}{2}$ and the relation (11) will look like:

$$\alpha = \frac{k_0(b-1)}{e(b+1)}\left(2 + \frac{\Delta E}{2k_0 T}\right). \tag{12}$$

here $b = \frac{u_n}{u_p}$, $\frac{\Delta E}{2} = \mu$.

From the gained relation (12) it is visible that quantity TEF for the natural semiconductor is spotted only by forbidden band breadth ΔE and a relation of mobility of charge carriers.

Expression (4) is gained, actually, in approach, when derivative of function much less than the function. It means that times much major, then warmth and charge relaxation times are considered. During too time presence of dependence of temperature and potential from co-ordinates specify in presence of interior lapse rates. Therefore, it is not absolutely correct use of an equilibrium distribution function as it corresponds to concept of an equilibrium state of local approach.

Consecutive viewing of transport of a charge and heat for the systems which are in lapse rates of temperature and potential can be spent a method of the nonequilibrium statistical operator [10].

The method of the nonequilibrium statistical operator allows to write down a uniform fashion the equations featuring a kinetics, taking into account a principle of impairment of correlations, and to gain expressions for relaxation times.

As method bottom the combined equations - the dilated equations Neumann's background serves:

$$\frac{\partial \rho}{\partial t} + \frac{i}{\hbar}[H, \rho(t)] = -\epsilon\left(\rho(t) - \rho_{eq}(t)\right) \tag{13}$$

where H - a Hamiltonian of system's, $\rho_{eq}(t)$ - the quasi-equilibrium statistical f operator, and ϵ – - the infinitesimal radiant providing irreversibility, which $\epsilon \to 0$ after thermodynamic transition.

Hamiltonian we will choose in a view [11]:

$$\mathrm{H} = \mathrm{H}_0 + \mathrm{H}_1 + \mathrm{H}_2, \tag{14}$$

$$\mathrm{H}_0 = \sum_k E_k a_k^+ a_k + \sum_m E_m a_m^+ a_m + \sum_{mm'} J_{mm'} a_{m'}^+ a_m + \sum_q \hbar\omega c_q^+ c_q,$$

$$\mathrm{H}_1 = \sum\nolimits_{kmq} \left(V_q + M_q a_m^+ a_m\right) e^{iqR_m} a_{k+q}^+ a_k$$

$$+ \frac{1}{2} \sum_{kk'q} \frac{4\pi e^2}{q^2} a_{k+q}^+ a_{k'} a_{k'-q}^+ a_k$$

$$+ \sum\nolimits_{kk'mq'} \left(R_{mkq} a_m^+ a_{k'} a_{k'-q}^+ a_k + R_{mkq}^* a_k^+ a_{k'} a_{k'-q}^+ a_m\right),$$

$$\mathrm{H}_2 = \sum_{\boldsymbol{kqm}} \mathrm{V}_{\boldsymbol{kqm}} \left(\boldsymbol{c}_{\boldsymbol{q}}^+ + \boldsymbol{c}_{-\boldsymbol{q}}\right) \left(a_{k+q}^+ a_k + a_m^+ a_{m'} + a_{k+q}^+ a_m\right),$$

Here $a_{m(k)}^+ \left(a_{m(k)}\right)$- operators of a creation (annihilation) of electrons corresponding localized m (nonlocalized - k) to states, $\boldsymbol{c}_{\boldsymbol{q}}^+ \left(\boldsymbol{c}_{-\boldsymbol{q}}\right)$ - operators of a creation (annihilation) of phonons with a wave vector q, $J_{mm'}$ - a matrix element of a jump between the localized states, V_q - potential of scatterers, $M_q \approx \frac{4\pi e^2}{q^2} \sum_q \Psi_0^*(p)\Psi_0(p+q)$, $R_{mkq} \approx \frac{4\pi e^2}{q^2} e^{i(k+q) \cdot R_m} \Psi_0(k+q)$, V_{kqm}- - a matrix elements the interaction electron–phonon, obvious expression for which is spotted by the concrete mechanism of dispersion.

The formal solution (13) we will write down in a view:

$$\rho(t) = \rho_{eq}(t) - \lim_{\varepsilon \to +0} \int_{-\infty}^{t} dt' e^{-\varepsilon(t-t')} e^{-\frac{i(t-t')\mathrm{H}}{\hbar}} \left(\frac{\partial \rho_{eq}(t')}{\partial t'} + + \frac{1}{i\hbar} \left[\rho_{eq}(t'), \mathrm{H}\right] \right) e^{\frac{i(t-t')\mathrm{H}}{\hbar}}. \tag{15}$$

Boundary conditions we will accept, in view of independence of a Hamiltonian of time, in a standard view [10]:

$$\mathrm{Tr}\left(\rho_{\mathrm{eq}} \mathrm{n_m}\right) = \mathrm{Tr}(\rho \mathrm{n_m}) = <\mathrm{a_m^+ a_m}>,$$

$$Tr\left(\rho_{eq} n_k\right) = Tr(\rho n_k) = <a_k^+ a_k> \tag{16}$$

At the initial moment of time $<a_{k,m}^+ a_{k,m}>_{t=0} = \left(e^{(E(\mathbf{k})-\mu)/kT} + 1\right)^{-1}$ - an equilibrium distribution function of electrons with temperature T.

Expressions for streams of a charge and heat we will choose in a view:

$$j(\boldsymbol{r}, t) = -\frac{2e}{(2\pi)^3} \int v_{\boldsymbol{k}} f(\mathbf{k}, \boldsymbol{r}, t) d\mathbf{k}, \tag{17}$$

$$W(\boldsymbol{r}, t) = \frac{2}{(2\pi)^3} \int (E(\mathbf{k}) - \mu) v_{\boldsymbol{k}} f(\mathbf{k}, \boldsymbol{r}, t) d\mathbf{k}, \tag{18}$$

where μ - chemical potential. It is necessary to score that in an equilibrium state when $f(\mathbf{k}, r, t) = f_0(\mathbf{k}, r,)$ – an equilibrium distribution function, streams are equal to zero. Thus, the problem solution consists in a finding and, after substitution, expressions for streams of a charge and warmth (17) and (18).

We choose a distribution function in a view:

$$f(\mathbf{k}, r, t) = <a_k^+ a_k> + <a_{m'}^+ a_m> + <a_k^+ a_m> + <a_m^+ a_k>, \tag{19}$$

The select of a distribution function in the form of (19) is caused by majority carriers of a charge taking into account possibility of transition of electrons from non-local in the localized states. Actually, it means possibility of an interference of states in basic one-particle states [11]:

$$\mathbf{1} = \sum_m |m><m| + \sum_k |k><k|$$

Such select is caused by definition of dynamic variables in method nonequilibrium statistical operator as these variables spot charge and warmth transport. At the expense of a phonon subsystem it is difficult to express transport of heat in the form of a stream. However, the shape of a Hamiltonian (14) allows to express $<c_q^+ c_q>$ through $<a_{k,m}^+ a_{k,m}>$ in the second order on a constant an interaction electron–phonon.

As it is known [11] to write down the transport equations relation performance is necessary:

$$[P_k, H_0] = \sum_l c_{kl} P_l$$

Using commutation relations of pairs of fermi-operators we will gain for a $P_{11} = a_k^+ a_k, P_{22} = a_{m'}^+ a_m, P_{12} = a_k^+ a_m, P_{21} = a_m^+ a_k$ relation:

$$\begin{aligned}
&[P_{11}, H_0] = 0; [P_{22}, H_0] = 0; \\
&[P_{12}, H_0] = -\sum_{m'} [(E_k - E_m) - J_{mm'}\delta_{mm'}]\, a_k^+ a_m; \\
&[P_{21}, H_0] = \sum_{m'} [(E_k - E_m) - J_{mm'}]\, a_m^+ a_k;
\end{aligned} \tag{20}$$

Thus from (20) it is visible that in processes of transport the dominant role is played by $<a_k^+ a_m>$, $<a_m^+ a_k>$, $<a_k^+ a_k>$ addends and serves as the tank having the temperature and chemical potential, depending on co-ordinates and time. Therefore, it is possible to spot from the kinetic equations and coefficients from (17) to (18).

Let us write down system of the kinetic equations for $<P_{12}>$ and $<P_{21}>$ in a view:

$$\begin{aligned}
&\frac{\partial <P_{12}>}{\partial t} - \frac{1}{i\hbar}\sum_{m'} [(E_k - E_m) - J_{mm'}\delta_{mm'}] <P_{12} \geq = \\
&= S_{12}^{(1)} + S_{12}^{(2)}; \\
&S_{12}^{(1)} = \frac{1}{i\hbar} <[\, P_{12}, H_{int}] >; \\
&S_{12}^{(2)} = -\frac{1}{\hbar^2}\int_{-\infty}^{t} dt' e^{\varepsilon(t'-t)} < \left[H_{int}, [H_{int}, P_{12}] + i\hbar\, P_{12}\frac{\partial S_{12}^{(1)}}{\partial <P_{12}>} \right] >,
\end{aligned} \tag{21}$$

Where $H_{int} = H_1 + H_2$. Similar equation registers and for $<P_{21}>$ to within transposition of coefficients.

Carrying out commutated in collisional members, we will gain the kinetic equations for $<P_{12}>$ and $<P_{21}>$ in the second order on force constants and quadratic on $<P_{12}>$, $<P_{21}>$.

Thus writing down the kinetic equation for and carrying out integration on p we will gain the Eq. (21) in which the right part is spotted through.

Without giving bulky expressions for streams, taking into account dispersion mechanisms, we will give dependences of relaxation times for the elementary case of transfer in the first order.

As the free length usually, I use various relaxation times enters into relations for mobilities. At high temperatures i.e. when energy of a phonon $\hbar\omega_0$ is much less than energy of an electron k_0T dispersion it is possible to consider elastic. Considering that at dispersion on ultrasonic phonons $\hbar\omega_0 < < k_0T$ also we will gain:

$$\tau = \frac{\sqrt{2}}{4\pi} \frac{Ma^3(\hbar\omega_0)^2 \mathrm{E}^{1/2}}{Z^2e^4m^{*1/2}k_0T}. \tag{22}$$

Free length

$$l = \tau V = \frac{V}{2\pi} \frac{M}{m^*} \left(\frac{\hbar\omega_0}{Ze^2/a}\right)^2 \frac{\mathrm{E}}{k_0T}, \tag{23}$$

Were V – velocity of an electron.

In case of low temperatures, $\hbar\omega_0 > > k_0T$, The electron–phonon interaction becomes inelastic. In this case processes of uptake of phonons are possible only, and to enter a relaxation time it is impossible. But in case of low temperatures if to consider a requirement $\hbar\omega_0 > > k_0T$, the majority of the electrons immersing energy of a phonon, transfer in an energy interval from $\hbar\omega_0$ to $2\hbar\omega_0$. Such electrons will almost instantaneous let out phonons, since the relation of probability of emission to probability of uptake $\frac{N_q+1}{N_q} \approx \exp \frac{\hbar\omega_0}{k_0T} > > 1$. As a result of such uptake energy of an electron does not change almost. It allows to view interaction of an electron with optical oscillations of a lattice at very low temperatures as elastic and to enter a relaxation time.

Calculation for a relaxation time gives view expression [12]:

$$\tau_{\text{оп}} = \frac{3\sqrt{2}}{4\pi} \frac{Ma^3(\hbar\omega_0)^{3/2}}{Z^2e^4m^{*1/2}} e^{\frac{\hbar\omega_0}{k_0T}}. \tag{24}$$

From (15) it is visible that at low temperatures the relaxation time on optical phonons does not depend on energy, and on temperature depends exponentially.

Free length

$$l = \tau V = \frac{3a}{2\pi} \frac{M}{m^*} \left(\frac{h\omega_0}{Ze^2/a}\right)^2 e^{\frac{h\omega_0}{k_0T}} \sqrt{\frac{\mathrm{E}}{k_0T}}. \tag{25}$$

It is necessary to score that because of presence $\exp\left(\frac{\hbar\omega_0}{k_0T}\right)$ free length always more than interatomic distance, i.e. $l >> a$.

From the presented analysis follows that charge relaxation times have essentially non-linear temperature dependence and the dispersion mechanism. All it demands the approach which is distinct from the traditional.

4. A method of the extended irreversible thermodynamics

In the standard approach of modeling of distribution of warmth in a thermoelement [4] as a rule, the classical equations of balance of transport of warmth are used:

$$q = \alpha jT - \frac{j^2 rL}{2} \tag{26}$$

where q - a specific heating capacity, T - temperature теплоотдающей mediums, α, r - TEF and a thermomaterial specific resistance, j - current density, L - thermobranch length.

Prominent feature of model (26) is lack of a time dependence q and temperature ρ. In case of use as a branch of a thermoelement of a material with NTCR [6], the temperature dependence is essential and demands the account at modeling of parameters of a thermoelement. The consecutive account of relaxation processes can lead to occurrence of waves of warmth and essentially changes character of distribution of heat at an initial stage of process.

Other feature of modeling is comparison of effects of model with experiment. Enough compound circuit of excitation and temperature measuring's is observationally used. The most attractive the plan with use as a radiant of warmth and a charge of an impulse of a current (**Figure 4**) looks.

In this case, measuring temperatures in points T_i, at excitation by a current impulse (point T_1) it is possible to realize the direct plan of distribution of warmth along a thermoelement branch.

As it was already specified earlier, use of classical thermodynamics invokes certain fundamental problems because of presence in system of a thermal cell of streams of heat and a charge. Therefore, the thermodynamic approach possesses intrinsic discrepancy. Such problems can be avoided, using the approach of extended irreversible thermodynamics [13]. Extended irreversible thermodynamics far from local balance uses as new explanatory variables dissipation streams, i.e. heat stream q, mass flux J and stress tensor P. Thus, in nonequilibrium system entropy S is function not only classical variable, but also dissipation streams:

$$S = S\{U(x,t), v(x,t), C(x,t), q(x,t), I(x,t)\}.$$

Introduction of streams as explanatory variables quite defensible from the physical point of view. Really, if in system there is any stream it means the directed

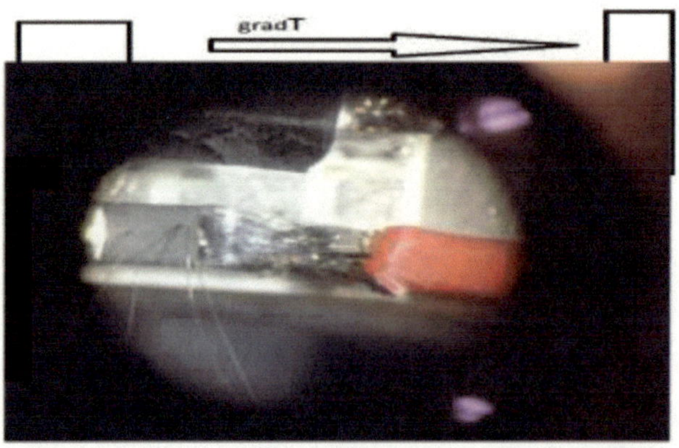

Figure 4.
Plan of modeling of a thermobranch.

locomotion of carriers of heat or mass. Hence, entropy which, as it is known, is a measure of affinity to an equilibrium state, specifies directions of locomotion of all system. Extended irreversible thermodynamics, in difference from classical is irreciprocal, introduces into processes time, as a variable that leads to the differential equations for dissipation streams of evolutionary (relaxation) phylum. In the elementary case relaxation times enter into such equations (Maksvella-Kataneo), in our case of heat and a charge.

Hence using methods of extended irreversible thermodynamics and the relaxation times, received from modeling representations about interactions in stuffs, it is possible to construct the consistent theory of the thermoelectric phenomena.

The initial combined equations can be written down in a view [13].

$$q + \tau_T \frac{\partial q}{\partial t} = -\lambda \nabla T + q_0(t, x) \tag{27}$$

$$\tau_e \frac{\partial i}{\partial t} = -(i - \sigma E') \tag{28}$$

where: $q_0(t,x)$ – source of heat, λ is the thermal conductivity coefficient, $E' = E - T\nabla(T^{-1}\mu_e)$, E is the electric field strength, μ_e is the chemical potential, τ_T, τ_e is the relaxation time of heat and charge, σ_e is conductivity, T is temperature.

Thus, the inclusion of dissipative flows in the series of independent variables leads to the fact that these flows are no longer determined by the gradient of the corresponding transfer potential, as in the classical local-equilibrium case, but they are solutions of the evolution Eqs. (27) and (28). These equations describe the process of relaxation of dissipative flows to their local-equilibrium values.

While analyzing the system of Eqs. (27) and (28), we use the following approximations. We assume that the coefficient of thermal conductivity and the relaxation time of the heat are constant and temperature is independent. Such assumption is correct in connection with the fact that the calculation of the heat relaxation time must be carried out taking into account the propagation of heat in the system. In other words, in the case of the heat propagation, the problem is self-consistent. Taking into account that the distribution of the heat and charge front may be considered in a single grain, we can assume that the spread of non-locality is rather weak, and the process describes the approach of the permanent τ_T and λ.

The τ_e - is an expression for the relaxation time of conduction electrons of a nondegenerate atomic semiconductor $\tau_e \sim \varepsilon^{-1/2}\, T^{-1}$ [12], where ε is an energy of the width order of the forbidden band of the semiconductor. Such an expression for the charge relaxation time is an approximation that has a temperature dependence. It is necessary to solve the kinetic equation for the charge propagation taking into account the dispersion law in the conduction mechanism [8]. However, in our case, such task is complicated by the fact that it is necessary to consider the flow of the charge along the grain surface. It complicates the solution of the kinetic equation, which must be solved taking into consideration the percolation flow model.

Let us converse (28) considering communication of reciprocity coefficients of Onsager's with phenomenological relations [13]:

$$\nabla T^{-1} = \frac{1}{\lambda T^2} q - \frac{\mu_e - \alpha T}{\lambda T^2} i; \tag{29}$$

$$E - \nabla \mu_e = \frac{\alpha}{\lambda} q - \left(\alpha \frac{\mu_e - \alpha T}{\lambda} - r \right) i; \tag{30}$$

Here Thomson's relation is used $\alpha T = -P$, were P – ccoefficient of Pelte. As a result of (28) transfers in

$$\frac{\tau_T}{\tau_e}\frac{\partial i}{\partial \tau} = -\frac{\mu_e - \alpha T}{r\lambda T}q + \frac{(\mu_e - \alpha T)^2}{r\lambda T}i \tag{31}$$

Time scale in (31) we will choose concerning relaxation times [10]. Here in expression (31) replacement is yielded $t/_{\tau_e} = {}^{\tau_T}/_{\tau_e}\tau; \tau = \frac{t}{\tau_T}$. Thus, the time dependence scale is spotted by the relation of relaxation times of warmth and a charge [12]. Conformity introduction between a return relaxation time and a thermal

conductivity ${\tau_T}^{-1'} \sim \frac{\lambda}{\lambda_q^2}$, were $\lambda_q = 2\,l$ – the doubled path length $\lambda_q = \tau V = \frac{3a}{\pi}\frac{M}{m^*}\left(\frac{\omega_0}{Ze^2/_a}\right)^2 e^{\frac{\omega_0}{k_0 T}}\sqrt{\frac{E}{k_0 T}}$, λ – thermal conductivity. $E' = E - T\nabla\left(T^{-1}\mu_e\right)$, τ_e – charge relaxation time, σ – conductivity, μ_e – chemical potential, E – electric intensity, T – temperature, ε– energy of the order of breadth of a forbidden band of a material.

Similarly, we will converse the Eqs. (27) and (31). We use a relation (29) and having presented $\lambda\nabla T = -\lambda T^2\left(\nabla T^{-1}\right)$, let us gain:

$$\tau_T\frac{\partial q}{\partial t} = -(\mu_e - \alpha T)i + q_0(t, x) \tag{32}$$

$$\frac{\tau_T}{\tau_e}\frac{\partial i}{\partial \tau} = \frac{\mu_e - \alpha T}{r}\frac{\nabla T}{T} \tag{33}$$

Further we use the law of conservation of energy and (32). It is as a result had:

$$\frac{1}{\tau_T}\frac{\partial^2 T}{\partial \tau^2} = -\frac{\alpha}{c_v\rho}i\nabla T + \frac{1}{c_v\rho}\frac{\partial}{\partial \tau}q_0(\tau, x) \tag{34}$$

were ρ – material density, c_v - specific heat capacity.

Thus, the Eqs. (33) and (34) feature distribution of a charge and temperature allocation to a thermoelement under the influence of a current impulse.

Warmth radiant we will choose in a view:

$$q_0(t, x) = ri^2 \tag{35}$$

I.e. a radiant is Joule heat. Thus, we consider that the current does not depend on co-ordinate and is spotted only by dependence from τ. Carrying out differentiation on τ and using (33), we will gain:

$$\frac{1}{\tau_T}\frac{\partial^2 T}{\partial \tau^2} = \frac{1}{c_v\rho}[\gamma\mu_e - (1+\gamma)\alpha T]i\frac{\nabla T}{T} \tag{36}$$

were $\gamma = \tau_e/\tau_T$.

As a result, we gain the combined equations featuring model of allocation of temperature along the sample at excitation by an impulse of a current.

$$\frac{\partial i}{\partial \tau} = \gamma\frac{\mu_e - \alpha T}{r}\frac{\nabla T}{T} \tag{37}$$

$$\frac{1}{\tau_T}\frac{\partial^2 T}{\partial \tau^2} = -\frac{i}{c_v\,\rho}\left[\alpha\nabla T - \gamma(\mu_e - \alpha T)\frac{\nabla T}{T}\right]$$

Prominent feature of model (37) is presence of the addends proportional to the relation of relaxation times of a charge and heat γ which acts as natural parameter little. The system (37) looks like a series development on, to the first order though at its deduction of any guesses about little it was not supposed. Thus, in the equation for change of temperature the addend of the zero order on γ is spotted by a thermoelectric stream of warmth. It shows that in the course of warmth distribution, to the initial moment of time, there is a heating at the expense of a current, as the most prompt.

Initial and boundary conditions for system we will choose in a view

$$i(0) = I; \tag{38}$$

$$T(0,0) = T_1; T(0,L) = T_2; \frac{\partial T}{\partial \tau}(0,0) = 0. \tag{39}$$

The predominant model of the conductivity of thermistors with NTCR is a model of hopping conductivity in the approximation of the "nonadiabatic" polaron of a small radius leading to the temperature dependence of conductivity [6]:

$$\sigma = \pi^{\frac{3}{2}} \frac{e^2 l^2 J^2 E^{-1/2}}{h(kT)^2} exp\,(-E/kT) \tag{40}$$

where: l is an effective hopping length, J is a parameter of jamping, E is an energy of hop activation, T is a temperature. Such nonlinear temperature dependence of electrical conductivity leads to a substantial nonequilibrium process of the heat and charge transfer in the branches of the thermoelement. It should be noted that in the case of a thermoelectric effect the process has a nonlocal character both in the coordinate and time. It is usually assumed that the chemical potential does not depend on temperature and is approximately equal to the Fermi energy. However, for a nondegenerate semiconductor with a temperature conductivity dependence (40), essentially nonlinear, it is necessary to consider the temperature dependence of the chemical potential [12], which has a logarithmic temperature dependence:

$$\mu_e = kTln\left[\frac{4}{3\sqrt{\pi}}\left(\frac{\varepsilon}{kT}\right)^{\frac{3}{2}}\right] < 0 \tag{41}$$

where k is a Boltzmann constant.

Thus, the problem of calculating the heat transfer in this system is nonstationary. To solve it, we assume the model to be one-dimensional, and dismiss the second-order terms in the temperature gradients. The initial and boundary conditions are assumed to be standard [14]. Such assumptions allow us to make a qualitative analysis of the nature of the propagation of heat and charge in the system. We examine the model at the distances of the grain size order. The generalization of the sample dimensions requires the establishment of an averaging procedure, which differs from the standard method i.e. the introduction of certain average or effective parameters requires additional considerations and cannot be carried out by simple averaging.

5. A construction and manufacturing of thermoelements

For manufacturing of structural thermoelements it is convenient to use the known production technology of chips-inductances.

Formation of structure of chips-thermoelements is carried out by level-by-level drawing ferrite and conductor layers. On a surface of a flat substrate superimpose thin (to several tens micron) a stratum of paste on the basis of the dielectric powder immixed with a binding material and solvents. On a surface of the dried stratum shape a printing expedient current-carrying (ferrite $NiMn_2O_4$ and $CoMn_2O_4$) drawing in the form of a semicoil, then superimpose the stratum coating 1/2 areas of preparation and leaving unclosed extremity of a semicoil. The following part of current-carrying drawing (semicoil) is superimposed so that the extremities of conductors were imposed against each other. Similar operations are iterated the necessary number of times that spending drawings in the form of semicoils were joined consistently, forming a spiral (**Figure 5**). Deficiencies of a "stage" expedient of manufacturing concern low reliability of switching of coils in places of transition of spending paste from the inferior plane on the upper.

Other, simpler plan of switching of the extremities of semicoils through a hole in the dielectric stratum (**Figure 6**) is developed also.

As linking of semicoils is yielded each time on half of interturn distance that twice reduces probability of formation of flaws on this critical site, thanks to the free diffluence of paste linking of coils trustier, structural structure trustier.

Structural thermoelements devices are made by a group expedient, i.e. simultaneously agglomerated a considerable quantity of devices on a substrate by the area 100×100 mm. After the termination of the making up the group package is slited on separate devices which are exposed to sintering.

Making of regenerative atmosphere allows to spend formation of a firm solution to furnaces with not compensated charge that leads to various type of conductivity depending on an initial composition.

The intermixture of gases CO - CO_2 can be gained two expedients. The first is grounded on interaction of gases CO_2 and the prosir-butanovoj of an intermixture, the second expedient - on restoration CO_2 at its gear transmission through a stratum of the heated coal. According to the first variant regenerative medium gained by conversion the prosir-butanovoj of an intermixture and CO_2. It occurs at temperature 800–1000°C. Formation of gases is accompanied by following responses:

$$
\begin{aligned}
&C_mH_{2m+2} \rightarrow mC + (m+1)\,H_2, C + H_2O \rightarrow \\
&\rightarrow CO + H_2, C + 2H_2O \rightarrow CO_2 + 2H_2, \\
&C + CO_2 \rightarrow 2CO
\end{aligned}
$$

The conducted examinations have shown that most full responses proceed at temperature 1000°C.

Advantage of the given method is that the rate of flux of gases is simply enough set with the help concentration of gas metter. However, enough high temperature is

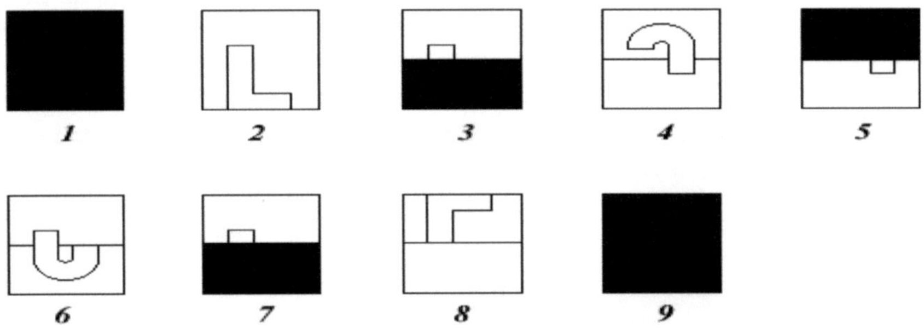

Figure 5.
The plan of the making up switching through a step.

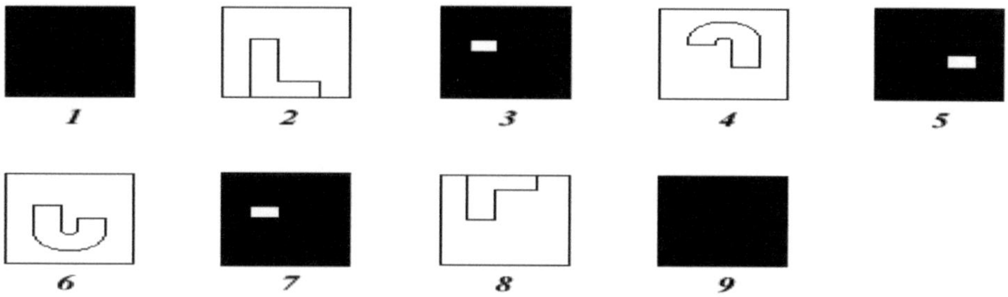

Figure 6.
The plan of the making up switching through a hole.

necessary for making of demanded atmosphere with an optimum exit of gases ~1000°C and the materials applied at manufacturing of the converter, should possess ability to work in regenerative medium (the quartz, special high-temperature grades of a steel). Besides, in this case takes place many-stage chemical responses that demands application of the special data units checking a composition of atmosphere on an exit of the converter.

On the second expedient a gas intermixture gained by gear transmission CO2 through the ceramic pipe filled with heated coal. There is an oxidizing of coal and restoration CO_2 on response:

$$CO_2 + C \rightarrow 2CO$$

In an equilibrium state at atmospheric pressure in a gas intermixture on a converter exit the following concentration CO in % contains: 2; 15; 58; 94; 99,3 accordingly at temperature 300, 500, 600, 700, 800°C. Advantage of this method is the low temperature at which there is a formation of regenerative medium. In our case it makes 650–700°C. Regulation of a composition of medium is carried out by change of temperature of the converter. It is necessary to carry necessity of a periodic fill of the converter to deficiencies coal.

Advantage of this method is the low temperature at which there is a formation of regenerative medium. In our case it makes 650–700°C. Regulation of a composition of medium is carried out by change of temperature of the converter. It is necessary to carry necessity of a periodic fill of the converter to deficiencies coal.

As builders it is possible to use restoration шпинелей $NiMn_2O_4$ and $CoMn_2O_4$ in the gas medium gained by conversion CO_2 and H_2O over carbon.

The offered procedure allows with sufficient repeatability to gain ceramic semiconductor materials with n and p conductivity types at simultaneous roasting.

6. Results and discussion

The results of numerical simulation are shown in **Figure 7**. The system analysis (37) was carried out for various relations τ_T/τ_e and the dimensionless time t/τ_T was used.

The result of the simulation is presented on **Figure 7a**, provided that the times of relaxation of heat and charge are close. In this case, the propagation of heat occurs almost simultaneously with the charge density. It has a character close to a solitary wave. Such result is quite obvious, since in this case Joule heat is released simultaneously with heat transfer and the increase of the charge current occurs with the velocity that is close to the velocity of propagation of heat V_1. A characteristic feature of such propagation is a formation of the wave on the length of the conductivity hop.

In the case when there is $\tau_T >> \tau_e$ (**Figure 7b**), the break of propagation front happens and the Joule heat wave V_2 outpaces the actual heat transfer wave due to the temperature gradient. Thus, in this case two waves are formed, which are spatially separated. At the same time, the relaxation of the heat does not occur during the hopping of the charge, and the system is in a locally nonequilibrium state. In other words, charge transfer generates a locally nonequilibrium state in which the charge flow is a fast variable.

In the case when there is $\tau_T << \tau_e$ (**Figure 7c**), the heat relaxation occurs faster than the charge transfer, and a heat propagation front coinciding with the charge transfer is formed. It should be noted here that the steepness of the front is determined by the mechanism of hopping conductivity and the approximations in the calculation. When there is more correct calculation there will be no gaps on the front.

The mode of heat transfer will be especially manifested in functional gradient materials [3], especially along the grain boundaries. By creating a regular structure with the required relaxation time ratios, it is possible to achieve the wave character of the heat transfer and charge transfer. It will allow to create devices that simultaneously measure and regulate the temperature.

Investigating movement of a charging and temperature wave it is possible to estimate the relation of times of a relaxation. It will allow, at qualitative level to draw certain conclusions about the carrying over mechanism. Such possibility is very actual for ceramic materials.

The first and obvious expedient of pinch of thermoelectric quality factor ZT of materials is optimization of their properties spotting thermoelectric efficiency. Such expedient allows to raise, though and is in most cases inappreciable, thermoelectric properties of traditional thermoelectric materials which are well studied and for which the physical analogues allowing purposefully to spot a direction of optimization of properties of materials are developed. As the basic directions of such optimization it is possible to ooze:

- Optimization of concentration of charge carriers;
- Optimization of breadth of a forbidden band;
- Optimization of a chemical compound of a material.

Optimum concentration of charge carriers (i.e., a select of an optimum level of a doping in the course of synthesis of a thermoelectric material) allows to provide the peak value of a thermoelectric quality factor. Physically, existence of optimum concentration of charge carriers is related by that at magnification of concentration the direct-current conductivity σ grows, and value termo-EDS, on the contrary, decreases. Dependence σ from concentration of charge carriers n is obvious and directly follows from direct-current conductivity definition.

Dependence termo-EDS from concentration of charge carriers is caused by a gas degeneracy of carriers at magnification of concentration of the item. For degenerated electronic gas Fermi level E_F gets to a conduction band (for the electronic semiconductor), and requirement E_f-E_c> kT, where E_C - energy of a bottom of conduction band, k - a Boltzmann constant is satisfied.

In this case energy and velocity of electrons are spotted by value of a Fermi level and practically do not depend on temperature. For such semiconductor, in the presence of a lapse rate of temperatures on its opposite extremities, streams of

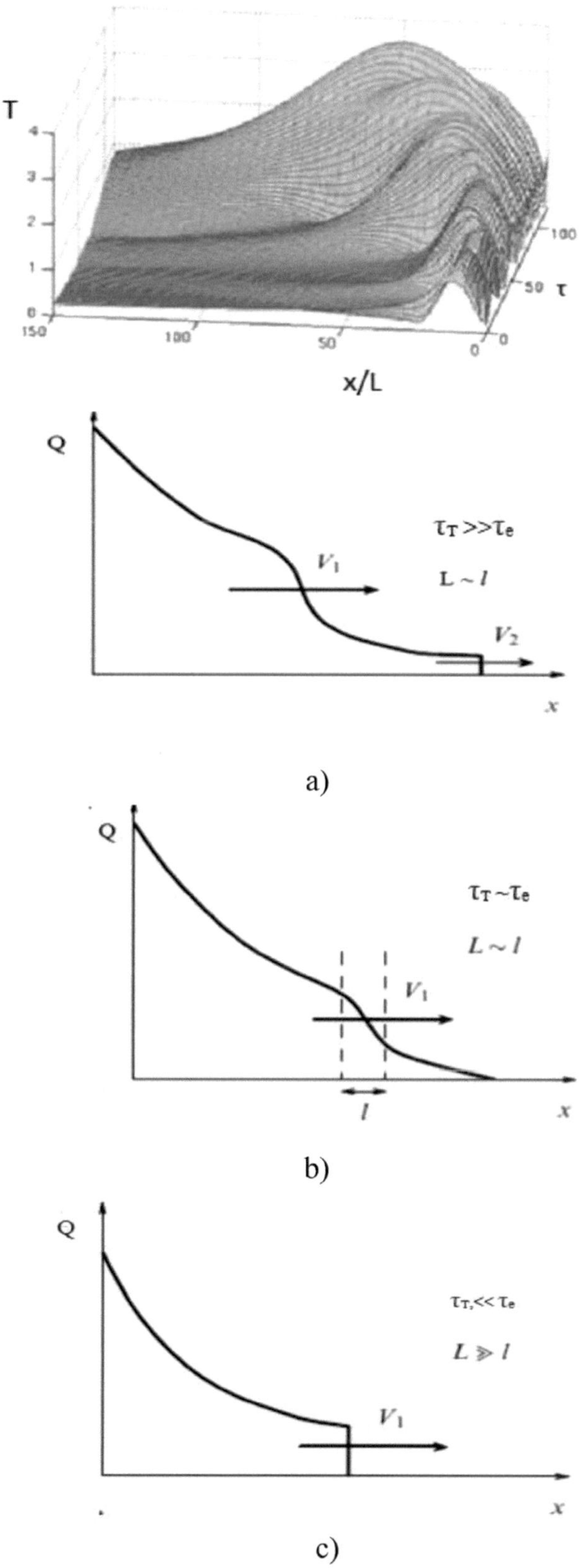

Figure 7.
Distribution of heat in a thermoelements for various ratios τ_T, τ_e, L.

electrons from the cold and hot extremities feebly differ, hence, volume termo-EDS will be inappreciable. Termo-EDS, and, hence, and a thermoelectric quality factor, it is possible to achieve pinch in semiconductors and semimetals in requirements when there is the strong degeneration, but concentration of charge carriers is great enough.

Optimization of breadth of forbidden band E_g is fundamental parameter of an electronic spectrum of the semiconductors which change allows to optimize their thermoelectric properties. As a result of a series of examinations [15] it has been shown that from the point of view of reception of a material with the best thermoelectric performances (for the unregenerate semiconductor) performance of following requirements is necessary:

- The forbidden band breadth should surpass essentially thermal energy κT,

- The Fermi level should settle down near to a bottom of an allowed band of majority carriers of a charge.

7. Conclusion

For the first time features of carrying over of warmth and a charge in semiconductor branches of a thermoelement are considered. The model of ceramic semiconductor branches is constructed of materials with negative temperature factor of resistance for thermoelements. Methods of the expanded irreversible thermodynamics spend modeling of distribution of temperature along the sample. On the basis of the spent modeling the technique of registration of impulses of temperature and definition of the relation of times of a relaxation in materials with negative temperature factor of resistance is offered. It is received that depending on a parity of times of a relaxation of processes of carrying over of a charge and warmth, various operating modes of a thermoelement can be realized.

It is offered to use by manufacture of thermobatteries the known production technology the chips-inductances. Feature of the specified technology is packing of separate elements in assemblage which then is divided into separate elements. Thus, instead of the semi-conductor materials containing rare and often ecologically dangerous materials, it is offered to use oxides ceramics. Roasting in regenerative atmosphere of the furnace allows to make layered thermoelements for one cycle of roasting.

Modeling of thermoelements from the ceramic materials possessing negative temperature in factor of resistance is spent.

Author details

Yuri Bokhan
Vitebsk Branch Educational Establishment, The Belarus State Academy of Communication, Vitebsk, Belarus

*Address all correspondence to: yuibokhan@gmail.com

References

[1] Rowe D.M. (ed.) CRC Handbook of Thermoelectrics. CRC. Boca Ration.- 1995. - 701 p.

[2] Rowe D.M. (ed.) Thermoelectrics handbook: macro to nano. CRC Press. Taylor & Francis group. 2006. - 954 p.

[3] Anatychuk, L.I. Current Status and some prospects of Thermoelectricity// Journal of Thermoelectricity. 2007. №2. P.7-20.

[4] Dmitriev, A.V., Zvyagin, I.P. Current trends in the physics of thermoelectric materials // Uspekhi Fizicheskikh Nauk. 2010. Vol. 180. №8. P. 821–838. DOI: 10.3367/UFNr.0180.201008b.0821

[5] Clarke, D.R. Oxide Thermoelectric Devices: A Major Opportunity for the Global Ceramics Community. // 5th International Congress on Ceramics, Beijing, August 2014.

[6] Feteira, A. Negative Temperature Coefficient Resistance (NTSR) Ceramic Thermistors: An Industrial Perspective. // J. Am. Ceram. Soc. 2009. vol.92. №5. p.967-983. DOI: 10.1111/ j.1551-2916.2009. 02990.x

[7] Koshi Takenaka Negative thermal expansion materials: technological key for control of thermal expansion // Sci. Technol. Adv. Mater. 2012. Vol. 13. P.1-10.DOI:10.1088/1468-6996/13/1/ 013001/

[8] Terasaki, I. High-temperature oxide thermoelectrics. //J.Appl. Phys. 2011.- 110.- 053705.

[9] Bokhan, Y. I., Varnava, A. A. Roasting of Ceramic Materials with the Negative Temperature Resistance Coefficient of Recovery Atmosphere. // Journal of Materials Sciences and Applications. 2018.Vol 4. No 3. p. 47-50. ISSN: 2381-1005 (Online)

[10] Röpke, G. Electrical Conductivity of Charged Particle Systems and Zubarev's Nonequilibrium Statistical Operator Method. Theor. Math. Phys.2018.194. p. 74–104. Doi:10.1134/ S0040577918010063

[11] Röpke, G. Electrical conductivity of a system of localized and delocalized electrons. Theor. Math. Phys.1981. 46. p. 184–190 Doi:10.1007/BF01030854.

[12] Veljko Zlatic, Rene Monnier Modern Theory of Thermoelectricity. Oxford. University Press. 2014. 289 p.

[13] Jou D., Casas-Vázquez J. and Lebon G. Extended Irreversible Thermodynamics, 3rd ed. Springer, Berlin. 2001.– 571 p.

[14] Bokhan, Y. I., Varnava, A. A. Thermoelectric ceramic element with a negative temperature factor of resistance //Journal of Thermoelectricity. 2018. No 1.p. 40-47.

[15] Melnikov, A.A., Kostishin, V.G., Alenkov, V.V. Dimensionless Model of a Thermoelectric Cooling Device Operating at Real Heat Transfer Conditions: Maximum Cooling Capacity Mode.// Journal of Electronic Materials, 2017.Vol. 46. No. 5. p.2737-2742. DOI: 10.1007/s11664-016-4952-0.

Chapter 8

Quantum Physical Interpretation of Thermoelectric Properties of Ruthenate Pyrochlores

Sepideh Akhbarifar

Abstract

Lead- and lead-yttrium ruthenate pyrochlores were synthesized and investigated for Seebeck coefficients, electrical- and thermal conductivity. Compounds $A_2B_2O_{6.5+z}$ with $0 \leq z < 0.5$ were defect pyrochlores and *p*-type conductors. The thermoelectric data were analyzed using quantum physical models to identify scattering mechanisms underlying electrical (σ) and thermal conductivity (κ) and to understand the temperature dependence of the Seebeck effect (S). In the metal-like lead ruthenates with different Pb:Ru ratios, σ (T) and the electronic thermal conductivity κ_e (T) were governed by 'electron impurity scattering', the lattice thermal conductivity κ_L (T) by the 3-phonon resistive process (Umklapp scattering). In the lead-yttrium ruthenate solid solutions ($Pb_{(2\text{-}x)}Y_xRu_2O_{(6.5\pm z)}$), a metal–insulator transition occurred at 0.2 moles of yttrium. On the metallic side (<0.2 moles Y) 'electron impurity scattering' prevailed. On the semiconductor/insulator side between x = 0.2 and x = 1.0 several mechanisms were equally likely. At x > 1.5 the Mott Variable Range Hopping mechanism was active. S (T) was discussed for Pb-Y-Ru pyrochlores in terms of the effect of minority carrier excitation at lower- and a broadening of the Fermi distribution at higher temperatures. The figures of merit of all of these pyrochlores were still small ($\leq 7.3 \times 10^{-3}$).

Keywords: Ruthenate pyrochlores, thermoelectricity, scattering mechanisms, metal–insulator transition, glass-like thermal conductivities

1. Introduction

The discovery of high thermoelectric performance of Na_xCoO_4 [1] has triggered renewed interest in oxide thermoelectric materials, though the figure of merit (zT) of this and other oxides is still too small for widespread application.

Pyrochlore is an oxide mineral [$(Na,Ca)_2Nb_2O_6$ (OH,F)] that forms brown to black, glassy octahedral crystals. In natural occurrences, the A and B atom sites can be occupied by many elements. Apart from naturally occurring compounds, over 500 compositions have been synthesized [2]. This wide-spread interest in pyrochlores is due to their large spectrum of properties, which include electronic, magnetic, electro-optic, piezoelectric, catalytic and more. The structure of pyrochlores is usually described by the topology and the geometric shape of coordination polyhedra. The structural formula is ${}^{VIII}A_2{}^{VI}B_2{}^{IV}X_6{}^{IV}Y$. The roman numerals show the coordination numbers. The crystal structure of pyrochlores is face-centered cubic (fcc). The space group is *Fd3m*, the lattice parameter is a = 0.9-1.3 nm. There are 8 formula units in the unit cell. X is oxygen (O^{2-}) and Y can be oxygen, hydroxyl, fluoride (O^{2-}, OH^-, F^-). The unit cell contains larger A (r_A = 0.087 – 0.151 nm) and smaller B cations (r_B = 0.040 – 0.078 nm) [3] surrounded by oxygens; in minerals by some OH^- and/or F^-. The B atoms are accommodated in distorted, corner-sharing BO_6 octahedra. A network of BO_6 octahedra forms the backbone of the structure [4]. The larger A atoms are located inside of slightly distorted hexagonal rings formed by six BO_6 octahedra. The structure can be regarded as two interpenetrating networks of B_2O_6 and A_2O' units. The pyrochlore structure tolerates vacancies on the A and O′ sites ('defect pyrochlores'), which can be represented by the general formula $A_{1\text{-}2}B_2X_6Y_{0\text{-}1}$. Work in this chapter deals with synthesizing lead ruthenate derivatives and lead-yttrium ruthenate solid solutions, measuring thermoelectric properties, and understanding the data based on the pyrochlore structure, A- and B-site occupancy, ligand and crystal fields as well as quantum-physical explanation of scattering mechanisms.

2. Electronic properties of ruthenate pyrochlores

Pyrochlores exhibit various electronic properties. Ruthenate pyrochlores with 4*d* transition elements on the B-site are of interest because their electronic properties can change when changing the A-site atom. The electrical conductivity is affected by the network of the corner-sharing RuO_6 octahedra, i.e., the B_2O_6 sublattice [4]. $Bi_2Ru_2O_7$ and $Pb_2Ru_2O_{6.5}$ show metal-like electrical conductivity to the lowest measured temperatures of a few degrees Kelvin [5, 6]. $Bi_2Pt_2O_7$ is an insulator [7]. $RE_2Ru_2O_7$ (rare earth RE = Pr to Lu), and $Y_2Ru_2O_7$ are Mott insulators with a spin-glass ground state [5, 8, 9]. Some pyrochlores show metal–insulator transitions (MIT), depending on temperature or pressure, for example $Hg_2Ru_2O_7$ and $Tl_2Ru_2O_7$ [10, 11]. $Tl_2Ru_2O_7$ shows a MIT at 120 K [9], likely due to the formation of a spin gap in the one dimensional Haldane chain [12]. Therefore, the electronic structure of a pyrochlore, especially near the Fermi level (E_F), must be affected by the element on the A-site via the A–O–Ru bond and the unoccupied O′ sites. Defect pyrochlores are cationic conductors, i.e., they are solid electrolytes. Some of them with 4*d* or 5*d* atoms on the B-site have been used as oxygen electrodes, because of their ionic (O^{2-}) and electronic conductivity [4].

There are at least two factors that may contribute to the electronic properties of Ru(IV) pyrochlores: 1) The Ru–O–Ru bond angle, which affects the Ru^{4+} t_{2g} band width and varies with the size of the A-site cation [13]. 2) Hybridization of unoccupied states of A-site cations (e.g., Tl, Bi, and Pb) with the Ru 4*d* states via the oxygen framework [14]. Changes of the Ru–O–Ru bond angle depend also on the Ru–O bond length, which is affected by the size of A and the resulting changes in orbital overlap and bandwidth [13, 15]. Respective studies have shown that pyrochlores with metallic behavior have greater Ru–O–Ru bond angles than that of insulating ruthenate pyrochlores [16]. There seem to be only small structural changes in $A_2Ru_2O_7$ pyrochlores when comparing metallic and semiconducting members. Electrical transport properties may just be positioned near the edge of localized and itinerant electrons. This positioning at the edge of the metal–insulator divide may help finding promising thermoelectric materials. High Seebeck coefficients are on the insulator side and potentially acceptable electrical conductivity on the metallic side, providing a balance in properties [4, 5].

Lee et al. [13] computed the band width of Ru t_{2g} block bands for various pyrochlore ruthenates using the extended Hückel tight binding method and showed that metallic phases have a wider bandwidth than semiconducting phases. The authors concluded that the metal-versus-insulator behavior of ruthenium pyrochlores can be explained in terms of the Mott-Hubbard mechanism of electron localization. Lee et al. [13] showed that there is a linear relationship between the ionic radius of the A cation and the Ru–O–Ru bond angle. I used the software *ATOMS V6.4.0* to calculate the Ru–O–Ru bond angle, which is 136.24°. The ionic radius for 8-coordinated Pb^{2+} is 0.129 nm. This is the highest bond angle and the largest ionic radius compared with the data shown in Figure 3a in [13]. The data confirm that Pb^{2+} follows as well the trendline established for cations on the A-site in ruthenate pyrochlores. Based on the high bond angle of 136.24°, $Pb_2Ru_2O_{6.5}$ can be expected to fall in the group of metallic ruthenate pyrochlore phases with a high Ru t_{2g} band width as shown by Lee et al. in Figure 3b in [13].

Lead ruthenate $Pb_2Ru_2O_{6.5}$ shows metal-like conductivity at room temperature. The ceramic has been used as catalyst for fuel cells, organic syntheses, and charge storage capacitors [4, 17]. As with all pyrochlores, the B atom, here Ru, is six-fold coordinated: each Ru atom is located at the center of a slightly distorted octahedron of equidistant oxygen atoms. Lead is VIII-coordinated with oxygen. $Pb_2Ru_2O_{6.5}$, or more precisely, $Pb_2Ru_2O_6O'_{0.5}$ forms an ordered defect-pyrochlore structure in which every other O′ site is empty [18]. The vacancies in the O′ sites are ordered. Following Hsu et al. [14] discussion, the metallic behavior of lead ruthenate is most likely due to the formation of an extended Pb 6*p* band overlapping with the Ru 4*d* band and is attributed to electron transport mainly within the octahedral network of BO_6 units, the backbone of the pyrochlore structure [4, 14, 19]. The fivefold Ru 4*d* levels are divided into an unoccupied e_g band ($\sim$ 2-5 eV above the Fermi energy E_F) and a partially occupied t_{2g} band ($\sim$ 1 eV below to 1 eV above E_F). The t_{2g} band is broad and mainly metallic due to covalent admixture with O 2*p* states. The Pb 6 *s* state is too deep ($\sim$ 8.5-10 eV below E_F) to be mixed with the Ru 4*d* states (the Fermi levels are in the same partially filled Ru 4*d* band). However, the unoccupied Pb 6*p* bands (only 5-9 eV above E_F) are close enough to mix with the Ru 4*d* states.

3. Quantum physical background

Following semiconductor physics, transport theory of metals, and degenerate semiconductors (parabolic band, energy-independent scattering approximation [20]) the Seebeck coefficient *S*, the electrical resistivity ρ [21], and the total thermal conductivity κ can be expressed as

$$\sigma = \frac{1}{\rho} = \pm en\mu \tag{1}$$

$$S = \frac{8\pi^2 k_B^2}{3eh^2} m^* T \left(\frac{\pi}{3n}\right)^{2/3} \tag{2}$$

$$\kappa = \kappa_e + \kappa_L \tag{3}$$

where *e* is the charge of a carrier, *n* is the carrier concentration ($1/m^3$), μ the carrier mobility ($m^2/V{\cdot}s$), and the plus or minus sign in Eq. (2) depends on the type of charge carrier, holes (+) or electrons (−); k_B is the Boltzmann constant, *h* is Planck's constant, *T* is the absolute temperature. The m^* in Eq. (2) refers to the effective mass of the charge carriers. Heavier carriers (higher m^*) will move more slowly. Hence, less mobility leads to lower electrical conductivity but a higher Seebeck coefficient (Eq. (2)). In other words, both, a smaller carrier concentration and a higher carrier's effective mass decrease electrical conductivity. The exact relationship between effective mass and carrier mobility depends on the electronic structure of a given material, on whether the material is isotropic or anisotropic and on scattering mechanisms [22].

The Wiedemann-Franz law (Eq. (4)) provides an expression for κ_e in Eq. (3):

$$\kappa_e = \sigma LT = en\mu LT \tag{4}$$

where σ is the electrical conductivity, L the Lorenz number, T is the absolute temperature. For metals and degenerate semiconductors L assumes the Sommerfeld value of 2.44×10^{-8} W·Ω/K^2. For non-degenerate, single parabolic band materials, and acoustic phonon scattering conditions, L drops to 1.49×10^{-8} W·Ω/K^2 [23]. The Wiedemann-Franz law is based on the assumption that free electrons (an electron gas) transport heat and electricity in metals, for which the total thermal conductivity is approximately equal to κ_e. The value of the Lorenz number varies among materials and depends, e.g., on band structure, on the position of the Fermi level, and on temperature; for semiconductors L relates to the carrier concentration [24]. The Lorenz number can vary, particularly with carrier concentration, e.g., in low-carrier concentration materials it can be about 20% lower than for metals [25]. Important deviations from the Wiedemann-Franz law are seen, e.g., in multi-band materials [26, 27], in nanowires [28], in superconductors [29], and in the presence of disorder [30]. The latter will be important for the materials investigated here. κ and σ are always determined experimentally. An accurate determination of κ_e is critical, since κ_L is often calculated from the difference between κ and κ_e (Eq. (3)).

Electronic transport processes can be analyzed using the density functional theory (DFT), electronic band-structure calculations, and the Boltzmann transport theory [31–33]. The Fermi-Dirac distribution function can be included in the computations to deal with temperature effects. To get a more complete description of electron-transport or electron scattering in a solid material, the energy (ε) dependence of the relaxation time of electrons near the Fermi energy is needed. To calculate the σ (T), S (T), and $\kappa_e(T)$ the Boltzmann transport equation framework is applied.

A scattering mechanism includes scattering from charge carriers and phonons. Scattering from charge carriers includes ionized impurities, acoustic phonons, the phonon deformation potential and scattering at crystal boundaries. Phonon scattering consists of phonon–phonon-, point-defect- and phonon-carrier scattering. To calculate the total scattering rate as the sum of the individual contributions Eq. (5), Mathiessen's rule is used:

$$\frac{1}{\tau_{\text{total}}} = \sum \frac{1}{\tau_i} \tag{5}$$

where τ_i is the relaxation time for each scattering mechanism and τ_{total} is total relaxation time. Note that because (ε) is integrated in the Boltzmann framework, the temperature dependence of conductivity may not be the sum of the temperature dependence of simple scattering processes. For example, if scattering process A has a relaxation time of $\tau_A(\varepsilon)$ and scattering process B has a relaxation time of $\tau_B(\varepsilon)$, then the total ε dependence will be

$$\frac{1}{\tau_{\text{t}}(\varepsilon)} = \frac{1}{\tau_{\text{A}}(\varepsilon)} + \frac{1}{\tau_{\text{B}}(\varepsilon)} \tag{6}$$

$$\tau_{\text{t}}(\varepsilon) = \frac{\tau_{\text{A}}(\varepsilon) + \tau_{\text{B}}(\varepsilon)}{\tau_{\text{A}}(\varepsilon)\tau_{\text{B}}(\varepsilon)} \neq \tau_{\text{A}}(\varepsilon) + \tau_{\text{B}}(\varepsilon) \tag{7}$$

As shown in Eq. (7), the combination of two scattering mechanism is not simply a summation, unless other approximations apply.

Table 1 shows that many different scattering mechanisms have been published. Here, I provide a list of examples of scattering mechanisms of electrical-, and thermal conductivities, as well as the Seebeck coefficient for one case. T_F is the Fermi temperature, and T_{BG} is the Bloch-Grüneisen temperature.

Umklapp or inharmonic scattering is present in crystalline materials at high temperatures and is the scattering of phonons by other phonons in three-phonon resistive processes. The significance of the process increases as the temperature increases, i.e., when more phonons with larger wave vectors are excited [38].

Scattering mechanism	Temperature	κ or S	σ
Electron-Impurity Scattering [34]	$T \ll T_F$	$\kappa_e \propto T$	$\sigma \propto T^{0}$
Electron-Impurity Scattering [34]	$T \gg T_F$	$\kappa_e \propto T^{1/2}$	$\sigma \propto T^{-1/2}$
Electron–Phonon Scattering [34]	$T \ll T_{BG}$	$\kappa_e \propto T^{-2}$	$\sigma \propto T^{-5}$
Electron–Phonon Scattering [34]	$T_F \gg T \gg T_{BG}$	$\kappa_e \propto T^{0}$	$\sigma \propto T^{-1}$
Electron–Phonon Scattering [34]	$T \gg T_F$	$\kappa_e \propto T^{-1/2}$	$\sigma \propto T^{-3/2}$
Umklapp Scattering [35]	High T Low T	$\kappa_L \propto T^{-1}$ $\kappa_L \propto T$	—
Electron–Electron Scattering [36]	Low T	—	$\sigma \propto T^{-2}$
Mott Variable Range Hopping [37]	Low to high T	$S \propto T^{1/2}$	$\sigma \propto e^{-(T_0/T)^{1/4}}$
Efros-Shklovskii Variable Range Hopping [37]	Low to high T	—	$\sigma \propto e^{-(T_0/T)^{1/2}}$
Nearest Neighbor Hopping [37]	Low to high T	—	$\sigma \propto e^{-(T_0/T)^{1}}$

Table 1.
Scattering mechanism of electrical-, thermal conductivities, and Seebeck coefficient.

In many semiconductors that show 'glass-like' behavior, there is a change in the conduction mechanism with temperature from thermally activated to 'Variable Range Hopping' (VRH) conduction [39, 40]. According to Mott, electron hopping between nearest neighbor sites is not always favored at low temperatures as the sites may be significantly different in energy. It is possible that electrons prefer to move to an energetically similar but more remote site. In this regime, the following conduction law is expected for the variation of the conductivity of glass-like disordered systems [41, 42]:

$$\sigma = \sigma_0 e^{-(T_0/T)^{\Upsilon}} \tag{8}$$

where σ_0 is the pre-exponential factor, T_0 is a characteristic temperature. The condition $0 < \Upsilon < 1$ is fulfilled, if the VRH mechanism dominates the conduction. Υ = ¼ corresponds to a Mott VRH, if the density of states around the Fermi level can be assumed to be constant; Υ = 1/2 correlates with an Efros-Shklovskii VRH, if there is a gap at the Fermi level [41, 42]; Υ = 1 agrees with a Nearest Neighbor Hopping (NNH) mechanism.

The most proper way to determine the accuracy of Υ in Eq. (8) is to analyze the electrical conductivity versus temperature by using the approach of Zabrodskii and Zinoveva [43] as follows. Let W be defined as

$$W(T) = \ln\left[\frac{d\ln\sigma(T)}{d\ln T}\right] \tag{9}$$

By inserting Eq. (8) into Eq. (9), $W(T)$ can be written as Eq. (10)

$$W(T) = \ln \Upsilon + \Upsilon \ln T_0 - \Upsilon \ln T \qquad (10)$$

$W(T)$ can also be used to determine metallic and insulating behavior. If the slope of ln $[W(T)]$ vs. ln T is negative, the material is an insulator. However, if the slope is positive, the material behaves like a metal [37]. Generally, in semiconductors (doped or un-doped) $\sigma(T)$ and $S(T)$ have opposite temperature dependencies. In Mott's Variable Range Hopping the electrical conductivity depends on temperature as $\sigma \propto e^{-(1/T)^{1/(d+1)}}$ and the Seebeck coefficient as $S \propto T^{(d-1)/(d+1)}$, where d is the dimensionality of the system [44]. In our work, the system is 3D; $\sigma(T)$ and $S(T)$ vary with temperature as $\sigma \propto e^{-(1/T)^{1/4}}$ and $S \propto T^{1/2}$.

4. Selecting and making lead- and lead-yttrium ruthenates

Substitutions of atoms on the A- and/or B-site in pyrochlore, here $Pb_2Ru_2O_{6.5}$, change the thermoelectric properties. I have selected and synthesized two sets of ruthenate pyrochlore compounds, one with variable Pb:Ru ratio, the other constitutes solid solutions of Pb- and of Y-ruthenate (**Table 2**). Changing the Pb:Ru ratio changes A and B site occupancy and the defect concentrations. For example, reducing the number of Pb^{2+} ions creates more vacancies in the A_2O' sublattice and changes the properties of the already existing oxygen vacancies, which are occupied by Pb $6s^2$ electron lone-pairs in $Pb_2Ru_2O_{6.5}$. Less than two Pb^{2+} ions mean less electrons for the vacancies. Reducing Ru should affect electrical conductivity, which is mainly due to the RuO_6 backbone structure of ruthenate pyrochlores.

Compounds*	Stoichiometric variations
$Pb_{(2+x)}Ru_{(2-x)}O_{(6.5\pm z)}$	x: 0.0, −0.3, 0.3, 0.5, 0.7, 0.9; $0 \le z \le 0.5$
$Pb_{(2-x)}Ru_2O_{(6.5\pm z)}$	x: 0.2; $0 \le z \le 0.5$
$Pb_{(2-x)}Y_xRu_2O_{(6.5\pm z)}$	x: 0.0, 0.1, 0.2, 0.4,1.0, 1.5, 2.0; $0 \le z \le 0.5$

**z was not measured; z = 0, known for $Pb_2Ru_2O_{6.5}$ and z = 0.5 for $Y_2Ru_2O_7$.*

Table 2.
Lead ruthenate derivatives and lead yttrium ruthenate solid solutions.

Partial Pb-Y substitutions have been studied and showed a transition from metal-like electrical conductivity to semiconducting, prior to becoming an insulating material [45]. Since Pb in $Pb_2Ru_2O_{6.5}$ causes the width of the Ru t_{2g} block band to be fairly wide and that of yttrium to be narrower [13], then yttrium is an interesting element when expecting property changes for partial substitutions. Obviously, $^{VIII}Y^{3+}$ with its much smaller ionic radius than $^{VIII}Pb^{2+}$ reduces the Ru–O–Ru bond angle [13], thereby making the increasingly Y-rich compound less metallic.

The pyrochlores were made as follows. Equally sized powders, about 10 μm in size were cold pressed into pellets, reacted at high temperature (1173 K to >1273 K), crushed, pressed again into pellets, sintered, cooled, and then measured. Only $Pb_2Ru_2O_{6.5}$ and $Y_2Ru_2O_7$ have been studied by others. The electrical conductivity of $Pb_2Ru_2O_{6.5}$ has been measured frequently, as shown in **Table 3**.

The results depended on the method of synthesis and fluctuated between 120 ± 30 S/cm (298 K) [46] and 4651.2 S/cm (300 K) [6]. Temperatures higher than 598 K have also been studied [17, 47].

Reference*	Temp. range (K)	Synthesis	298-300 K	323 K	373 K	423 K	473 K	523 K	573 K
Tachibana et al. [6]	1 < T < 300	Single crystal	4651	—	—	—	—	—	—
Mayer-von Kürthy [7]	77 < T < 300	Solid state	2128	—	—	—	—	—	—
Takeda et al. [17]	298 < T < 1173	Solid state	631	513	479	447	417	389	251
Parrondo et al. [46]	298 < T < 598	Alkaline solution	120 ± 30	—	—	—	—	—	—
Song and Lee [47]	373 < T < 1073	Co-precipitation process	—	—	3490	3375	3250	3100	3150
Akhbarifar et al. [48]	298 < T < 598	Solid state	2137	2037	1902	1781	1702	1638	1579
Sleight [49]	4.2 < T < 298	Solid state	2000	—	—	—	—	—	—
Kobayashi et al. [50]	15 < T < 300	Solid state	1000	—	—	—	—	—	—

**Except Perrondo et al., all authors measured σ(T) with DC 4-probe technique.*

Table 3.
Known temperature dependences of electrical conductivity σ(T) of lead ruthenate.

5. Lead ruthenate derivatives

X-ray powder diffraction measurements and phase identification were performed. Search-match routines in *JADE9* software (Materials Data. Inc.) were used. The powder patterns were calibrated using a corundum (α-Al_2O_3) standard. X-ray powder diffraction patterns of lead ruthenate ($Pb_2Ru_2O_{6.5}$) and six derivatives are shown in **Figure 1**. All *d*-spacing and intensities match the isometric pyrochlore crystal structure. Lattice constants a_0 were calculated for all samples. The lattice constant of pure lead ruthenate (a_0 = 1.0257 nm) was in good agreement

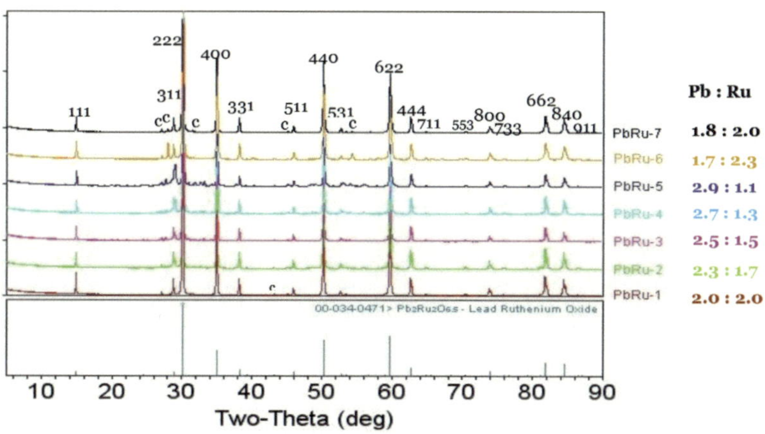

Figure 1.
X-ray diffraction patterns of lead ruthenate pyrochlore [51] and six derivatives (below: Reference pattern 00-034-0471 of $Pb_2Ru_2O_{6.5}$; 'c' = unidentified contaminant).

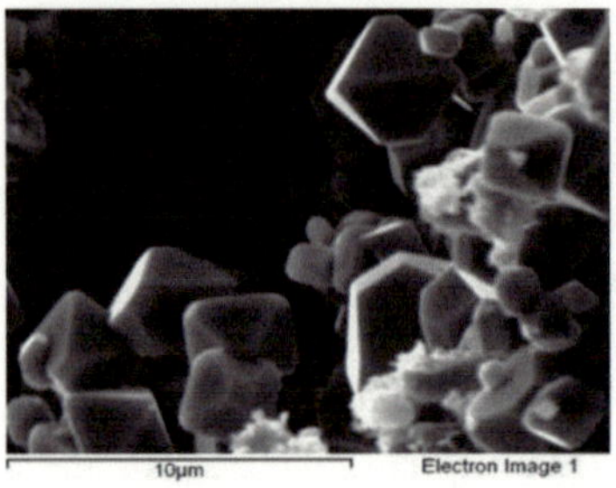

Figure 2.
SEM image showing octahedral crystals of our $Pb_2Ru_2O_{6.5}$.

with Jade reference # 00-034-0471 by Horowitz et al. [52] with a value of 1.0252 nm.

Substitutions affect a_0, but the isometric unit cell is maintained in all cases. A closer look at the X-ray spectra at 2θ = 28° showed that traces of unreacted RuO_2 may be present, particularly in Pb:Ru = 1.7:2.3. The yield of all synthesized pyrochlores is close to 100%.

Solid-state synthesis of pyrochlores involved a series of manipulations, which introduced ZrO_2 from milling containers, SiO_2 and Al_2O_3 from glass mortars and pestles. The impurities were analyzed by semi-quantitative X-ray fluorescence and amounted to ≈ 1 wt.%. When present, unreacted RuO_2 amounted to ≈ 1 wt.% as well. Suggesting an overall purity of our pyrochlores of at least 98 wt.%. **Figure 2** shows a secondary electron image of octahedral $Pb_2Ru_2O_{6.5}$ crystals.

5.1 Electrical- and thermal conductivity and Seebeck coefficients of lead ruthenate derivatives

In this section I present and discuss thermoelectric properties of derivatives of lead ruthenate ($Pb_{(2+x)}Ru_{(2-x)}O_{(6.5\pm z)}$ and $Pb_{(2-x)}Ru_2O_{(6.5\pm z)}$), **Table 2**). Comparisons are made with the published results of lead ruthenate [48].

Figure 3a shows electrical conductivity as a function of temperature for lead ruthenate derivatives. All conductivities decrease with increasing temperature, suggesting metal-like performance of all ceramics. The lead ruthenate derivative with a deficiency of lead, i.e., $Pb_{1.8}Ru_2O_{6.5\pm z}$ exhibits the highest electrical conductivity. The lowest conductivity was seen for the derivative with an excess of lead and a deficiency of ruthenium, i.e., $Pb_{2.9}Ru_{1.1}O_{6.5\pm z}$. The oxygen content of the derivatives was not measured.

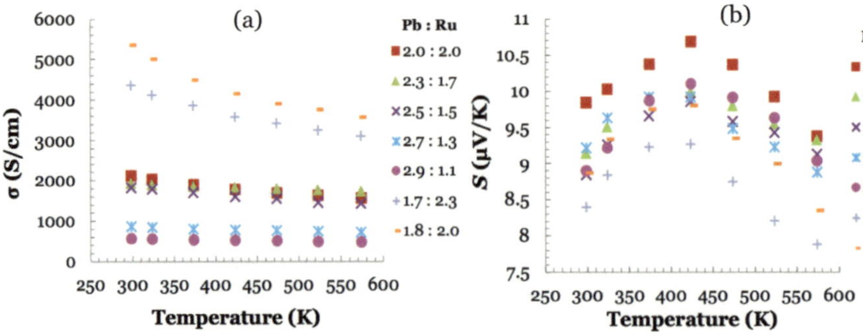

Figure 3.
(a) Electrical conductivity and (b) Seebeck coefficients of lead ruthenate derivatives.

Electrical conductivity is attributed essentially to transport of electrons in and through BO_6 octahedra, the backbone of the pyrochlore structure [53, 54]. The Pb $6s^2$ electron lone-pair is highly localized in defect pyrochlores like lead ruthenate with oxygen deficiency (O′ vacancies) in the sublattice. Hence, the electron lone-pair is stereo-chemically active. This results in the formation of ordered oxygen vacancies and the displacement of Pb closer to the vacancy [14]. In this way, the pyrochlore structure is stabilized vis-à-vis the perovskite structure [18]. Pb may be present in lead ruthenates in two valence states as $Pb^{2+}{}_{1.5}Ru^{4+}{}_{2}Pb^{4+}{}_{0.5}O_{6.5}$ [9]. Alternatively, Ru could be present in two valence states as Ru^{4+} and Ru^{5+}, i.e., $Pb_2Ru^{4+}{}_{1.0}Ru^{5+}{}_{1.0}O_{6.5}$. Even with some Pb^{4+} present, the pyrochlore structure would still be sufficiently stabilized by the Pb^{2+} $6s^2$ electron lone-pair.

Removal of Ru, i.e., decreasing the overlap of the Pb $6p$ with the Ru $4d$ bond, should result in a decreasing electrical conductivity, relative to pure lead ruthenate, which is confirmed by the experimental results (B-site stoichiometry <2). For charge balance reasons, it is assumed that Pb enters the B-site as Pb^{4+}, which also means absence of some Pb $6p$ electrons. Decreasing Pb^{2+} on the A-sites (x = 1.7) without substitutions increased the electrical conductivity significantly, which may be the result of creating more oxygen vacancies and vacancies in the A_2O' sublattice. The relatively strong increase of electrical conductivity due to increasing Ru to 2.3 at the expense of Pb to 1.7 is less easy to explain. The replacement of 2 Pb^{2+} by one Ru^{4+} on the A site would cause an A-site vacancy, which could increase electrical conductivity. Ru^{4+} is smaller than Pb^{2+}. This should cause rattling of Ru, which would increase phonon scattering and thus decrease thermal conductivity. This will be discussed further in context with thermal conductivity and scattering mechanisms.

Figure 3b shows that the Seebeck coefficients go through a maximum near 423 K for all lead ruthenate derivatives. The dependence of the Seebeck effect on the composition of these pyrochlores is not very pronounced. The highest and lowest coefficients deviate from each other by about 15%, independent of temperature. A surprising finding was that $S(T)$ increased initially with increasing temperature and then decreased with further increasing temperature. Since $S(T)$ is positive the majority of charge carriers are holes, i.e., lead ruthenate pyrochlore and its derivatives exhibit p-type conductivity. The kind of temperature dependence of S found for these pyrochlores has been observed in other systems as well. For example, $Y_{2-x}Bi_xRu_2O_7$ show a broad maximum at 170 K for x = 1.6 [9], but the authors gave no explanation. Paschen [51] observed a maximum of $S(T)$ in a clathrate-like material, i.e., in $Eu_3Pd_{20}Ge_6$, at about 120 K. The author attributed the maximum to valence fluctuations of Eu (3^+ and 4^+). Sidharth et al. [55] measured a broad maximum of $S(T)$ for tin chalcogenides, $SnSe_{1-x}Te_x$ between 400 K and 600 K. These authors discuss their findings in terms of the influence of carrier-phonon-, carrier-carrier scattering and conductivity-limiting bipolar conduction at high temperature. Tse et al. [56] reported a maximum of S for Na_xSi_{46} clathrates at x = 16 for 200 K. Certain clathrates are metallic, but their thermal conductivity resembles that of glass-like solids, as with lead ruthenate [48] and the derivatives studied here. Tse et al. [56] ascribed the trend of $S(T)$ as a function of sodium content to the band profile of the silicon framework. The rise of $S(T)$ with increasing temperature to a maximum and the decrease with further increasing temperature was qualitatively reproduced by raising the Fermi level. A maximum in the temperature dependence of $S(T)$ was also seen in bismuth metal thin films. For a 33 nm thick film, the maximum occurred near 425 K [57]. The authors of [57] discussed the phenomenon by taking into account the temperature dependence of the Fermi energy, which cannot be neglected because bismuth has a low Fermi energy. Hence, increasing and decreasing Seebeck coefficients have been observed in a variety of materials and in different temperature ranges. Explanations of respective data for different materials vary.

The Seebeck coefficient of these *p*-type pyrochlores is inversely proportional to the carrier concentration (Eq. (2)). Before the onset of intra-band minority carrier (electron) excitation at lower temperatures, $S(T)$ increases with T (**Figure 3b**), confirming metallic behavior of all pyrochlores. At higher temperatures the Fermi distribution broadens, which leads to an exponential increase in minority electrons, due to thermal excitation and resulting in a reduction of the Seebeck coefficient [58]. The Seebeck coefficient reaches a maximum, in this case at 423 K (**Figure 3b**) for all derivatives.

Figure 4 shows the results of thermal conductivity measurements of all lead ruthenate pyrochlores. Thermal conductivity increases with temperature and was found to have increased in all derivatives above that of lead ruthenate. Note that the compounds 1.7:2.3 and 1.8:2.0 showed the highest electrical conductivities (**Figure 3a**), while their thermal conductivities are least affected (**Figure 4**). This will be addressed in the next section. I noticed an increase of the total thermal conductivity with temperature in the crystalline lead ruthenate and its derivatives, which is typical of glass-like materials. A glass-like behavior of all these pyrochlores is likely due to an electronic contribution from an increasing concentration of minority electrons with temperature, which also contributed to the downturn of the Seebeck coefficients.

The presence of structural defects and a large unit cell create pronounced anharmonicity, which lowers the thermal conductivity [59]. $Pb_2Ru_2O_{6.5}$ possesses intrinsic disorder characteristics, i.e., vacancies in the A_2O' sublattice and stereochemically active $6s^2$ electron lone-pairs on Pb. Similar defects can be expected to prevail in our derivatives. These defects cause a glass-like temperature dependence of thermal conductivity of our lead ruthenate derivatives. This was also observed, e.g., in chalcogenides, clathrates, and yttrium-stabilized zirconia (YSZ) [38], and materials with intrinsic disorder [60–63]. The thermal conductivity of lead ruthenate is close to that of vitreous silica [48], which is 1.93 W/m·K at 373 K. With increasing temperature, the 3-phonon resistive process (Umklapp scattering) and excitation of more phonons with larger wave vectors become increasingly important [64], thereby lowering the lattice thermal conductivity of a crystal [59].

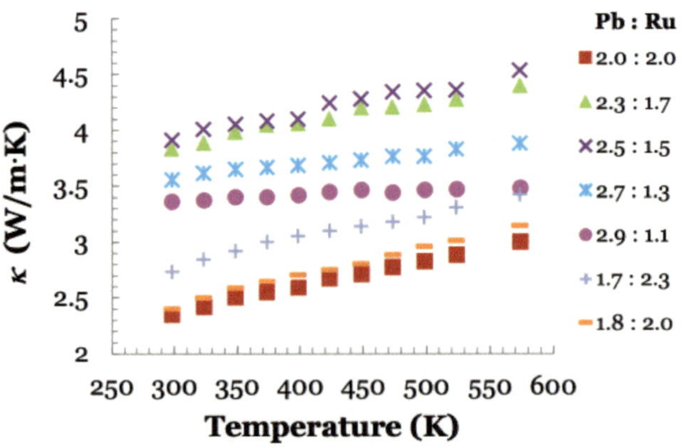

Figure 4.
Thermal conductivity of lead ruthenate derivatives.

5.2 Scattering mechanisms in lead ruthenate derivatives

The measured thermal conductivity κ is the sum of electronic κ_e and lattice thermal conductivity κ_L (Eq. (3)). The electronic and lattice contributions are plotted in **Figure 5a** and **b**, respectively. The electronic thermal conductivity increased with temperature and decreased with increasing Pb content. The Pb:Ru = 1.8:2.0 and the Pb:Ru = 1.7:2.3 pyrochlores, respectively, have the highest electronic thermal conductivities. With one exception, the lattice thermal conductivities are almost independent of temperature but vary with composition. Partial removal of Pb from $Pb_2Ru_2O_{6.5}$ lowers the lattice thermal conductivity. Using the Lorenz number of 2.44×10^{-8} WΩ/K^2 for metals to calculate κ_e yielded negative values for κ_L, which is meaningless. To avoid this, I used 1.49×10^{-8} WΩ/K^2, which has been applied for non-degenerate, single parabolic band materials, and acoustic phonon scattering conditions [62].

As expected from electrical conductivities (**Figure 3a**), the compounds with a lead deficiency show κ_e values higher than in pure lead ruthenate (**Figure 5b**). However, the κ_L values are the lowest of all. These compounds are presumably the ones with the highest defect concentrations of all lead ruthenate derivatives, which are assumed to increase phonon scattering and thus lower κ_L.

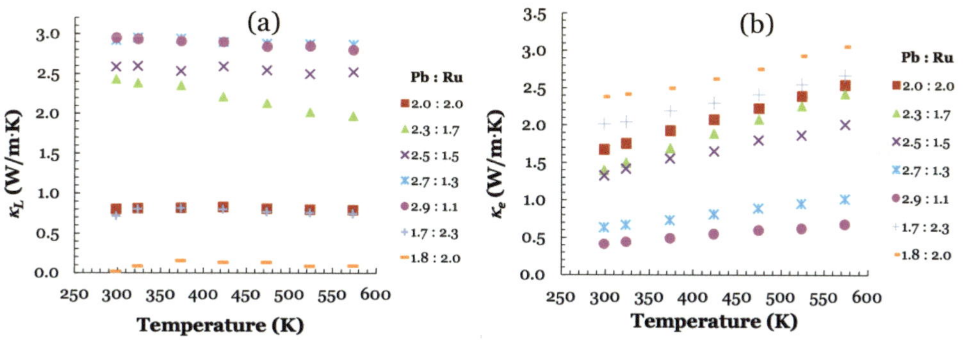

Figure 5.
(a) Lattice- and (b) electronic thermal conductivity of lead ruthenate derivatives.

The electrical and thermal conductivity data of lead ruthenate derivatives have been analyzed for underlying scattering mechanisms. The temperature dependence of the scattering mechanisms listed in **Table 1** were least-squares-fitted to the experimental data. The goodness of the fits was analyzed by using the reduced chi-squared statistic (χ^2/DOF) (Eq. (11)) and the best fit was assumed to have identified the most likely scattering process.

$$\chi^2/DOF = \frac{Sum\ of\ squared\ errors}{DOF} \tag{11}$$

where DOF is the number of degrees of freedom, which is the number of data points minus the number of fit parameters. For all seven synthesized pyrochlores, the best fits for $\sigma(T)$, $\kappa_e(T)$, and $\kappa_L(T)$ were $T^{-1/2}$, $T^{1/2}$, and T^{-1}, respectively. All fits for all pyrochlores pointed at the same underlying scattering mechanisms. A graphical representation of these dependencies can be found in [48] for pure lead ruthenate. According to **Table 1**, the underlying scattering mechanism is 'electron impurity scattering'. Since this mechanism prevails only at $T \gg T_F$ (**Table 1**), the excellent match between data and the fits suggests that the Fermi temperature T_F of lead ruthenate and its derivatives is below room temperature. Other joint scaling shapes rule out other combinations of dominant scattering mechanisms. This is evidence that traditional electron-acoustic phonon scattering is suppressed and thus most scattering is due to intrinsic disorder and impurities.

The lattice thermal conductivity $\kappa_L(T)$ of lead ruthenate and all derivatives varies with T^{-1}. This dependence suggests that the 3-phonon resistive process

(Umklapp scattering, **Table 1**) is responsible for the decrease of $\kappa_L(T)$ in all compositions. This finding supports the electron-impurity scattering mechanism for $\sigma(T)$ and $\kappa_e(T)$ (**Table 1**) [64], because transport properties are sensitive to structural disorder such as site substitutions, vacancies and localized impurities, all of which are present in these pyrochlores at different concentrations. These defects can lower the thermal conductivity to glass-like behavior [59]. In this regard, compounds with stereochemically active electron lone-pairs associated with constituent atoms (here Pb^{2+} on the pyrochlores' A-site) have attracted significant attention. Other examples are Cu_3SbSe_3, Ag_SbSe_2, and other chalcogenides. The 3-phonon resistive process will be addressed again in Section 6.2. The same causes as here hold there for Umklapp scattering for lead-yttrium ruthenate solid solutions. The electronic contribution $\kappa_e(T)$ to the total thermal conductivity $\kappa(T)$ increased with increasing temperature and was estimated to be about 72% at 25°C and 85% at 300°C, respectively. The total thermal conductivity $\kappa(T)$ results mainly from electronic thermal conductivity.

6. Lead yttrium ruthenate solid solutions

We have reported recently [45] on electrical conductivity and Seebeck coefficients of lead yttrium ruthenate solid solutions. The focus of that investigation was on finding and explaining a metal insulator transition in a suite of Pb-Y ruthenate solid solutions between the metal-like paramagnetic $Pb_2Ru_2O_{6.5}$ and the antiferromagnetic Mott insulator $Y_2Ru_2O_7$. Only 10 mol% of Pb needed to be replaced by yttrium to reach the point of transition from a metal-like to a semiconducting ceramic. Following the Mott-Hubbard model, yttrium opens the Mott-Hubbard gap and fills the lower Mott-Hubbard band with localized t_{2g} electrons thereby changing the mechanism of electron transport [45].

In this section I report on the hitherto unknown thermal properties of lead yttrium ruthenates and analyze these together with our previously published measurements on electrical conductivity and the Seebeck effect [45] in terms of quantum-physical scattering.

Structure and purity of all samples (**Table 2**) have been determined by X-ray diffraction and X-ray fluorescence analysis, respectively. The X-ray diffraction patterns of pure lead and yttrium ruthenate are known (Jade reference # 00-034-0471, a_0 = 1.0252 nm and # 00-028-1456, a_0 = 1.0139 nm, respectively). The solid solutions are all of isomeric structure and follow Vegard's law, i.e., the unit cell parameter decreases linearly with increasing concentration of Y^{3+} (r = 0.1019 nm), which is smaller than Pb^{2+} (r = 0.129 nm) [45].

6.1 Thermal conductivity of lead yttrium ruthenate solid solutions

The results of thermal conductivity measurements are shown in **Figure 6**. Thermal conductivity κ increases with temperature. Pure lead ruthenate has the highest thermal conductivity (2.35 W/m·K) of the pyrochlores at all temperatures [48]. As the Y concentration increases, κ and its temperature dependence decrease. Pure yttrium ruthenate has the lowest thermal conductivity.

κ increases with temperature up to an yttrium concentration of x = 1.0. At and above x = 1.5, κ is very low and nearly temperature independent. These pyrochlores are insulators as indicated by low electrical conductivity as well [45]. The increase of κ with increasing T, and the low values of κ are typical of glass-like behavior, although all these pyrochlores are crystalline. As with lead ruthenate derivatives, glass-like behavior is probably due to a contribution to κ_e of an increasing concentration of minority electrons with increasing temperature, which also contributed to the decrease of the Seebeck coefficient for x = 0 to 1.5 [45].

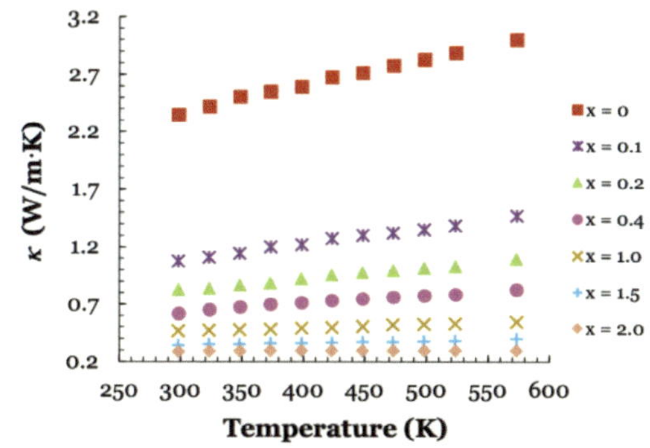

Figure 6.
Thermal conductivity of Pb- and Y-ruthenate and solid solutions.

6.2 Scattering mechanisms in lead yttrium ruthenate solid solutions

The electronic thermal conductivity κ_e was calculated with Eq. (4), the lattice thermal conductivity κ_L with Eq. (3) and plots are shown in **Figure 7a** and **b**.

The electronic thermal conductivity κ_e (**Figure 7a**) decreases with increasing Y content but increases with temperature. The slope of the temperature dependence of κ_e decreases with increasing concentration of yttrium. **Figure 7a** (lower diagram) contains the data for the two insulators $Pb_{(2-x)}Y_xRu_2O_{(6.5+z)}$ with x = 1.5 and 2.0 on a different scale, to show that there is still a temperature dependence of κ_e. $Y_2Ru_2O_7$ shows the lowest electronic thermal conductivity.

The lattice thermal conductivity κ_L decreases with increasing temperature for $x \leq 0.2$ (**Figure 7b**). Between x = 0.4 and x = 2.0 κ_L becomes temperature independent. κ_L increases initially with temperature in pure lead ruthenate (x = 0).

In this section electrical- and thermal conductivity and Seebeck coefficients of the pyrochlores $Pb_{(2-x)}Y_xRu_2O_{(6.5+z)}$ ($0 \leq x \leq 2$, $0 \leq z \leq 0.5$) will be analyzed for the underlying scattering mechanisms. To do so, the temperature dependence of the scattering mechanisms listed in **Table 1** were fitted to the experimental data. The goodness of all fits was analyzed by using Eq. (11), which yielded the most likely scattering mechanisms.

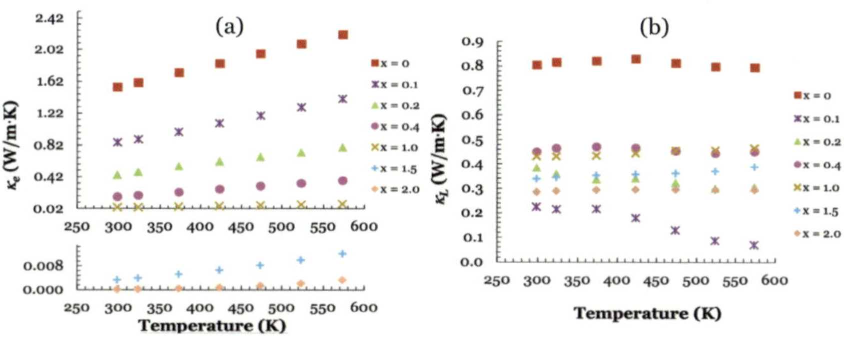

Figure 7.
(a) Electronic- and (b) lattice thermal conductivity of Pb- and Y-ruthenate and solid solutions.

As shown in **Table 1**, for most scattering mechanisms at least two properties, for example, $\sigma(T)$ and $\kappa_e(T)$ must be fitted together to show the respective temperature dependencies. For $Pb_{(2-x)}Y_xRu_2O_{(6.5+z)}$ with x = 0 and 0.1 the best fits of electrical conductivity data $\sigma(T)$ and of the electronic component of thermal conductivity $\kappa_e(T)$ vary with $T^{-1/2}$ and $T^{1/2}$, respectively (see **Figure 8a** and **b**). According to **Table 1** the underlying scattering mechanism is 'electron-impurity scattering' at $T \gg T_F$ [37]. The lattice thermal conductivity $\kappa_L(T)$ varies with T^{-1} suggesting that the 3-phonon resistive process (Umklapp scattering, **Table 1**) is responsible for the decrease of $\kappa_L(T)$ in the pyrochlores with x = 0 and x = 0.1 moles of yttrium. This finding supports the electron-impurity scattering mechanism for $\sigma(T)$ and $\kappa_e(T)$ (**Table 1**) [41]. The experimental data and the least squares fits are shown in **Figure 8a–c**.

For $Pb_{(2-x)}Y_xRu_2O_{(6.5+z)}$ with x = 0.2 and x = 0.4 the best fits of the electrical conductivity data $\sigma(T)$ and of the electronic component of thermal conductivity $\kappa_e(T)$ did not allow for an unambiguous determination of the scattering mechanism. The goodness of two fits was practically identical. Two variations of $\sigma(T)$ and $\kappa_e(T)$ with temperature are equally likely: $\sigma(T)$ varies with $T^{-1/2}$ or with T^{-1} and $\kappa_e(T)$ varies with $T^{1/2}$ or with T^{0}. Based on **Table 1** the underlying scattering mechanisms are 'electron-impurity scattering' at $T \ll T_F$ provided that $\sigma(T)$ varies with $T^{-1/2}$ and $\kappa_e(T)$ with $T^{1/2}$. If $\sigma(T)$ varies with T^{-1} and $\kappa_e(T)$ with T^{0} then 'electron–phonon scattering' at $T_{BG} \ll T \ll T_F$, is the mechanism (**Table 1**) [37]. The lattice thermal conductivity $\kappa_L(T)$ varies with T^{-1} implying Umklapp scattering and supports the electron-impurity scattering mechanism. The actual scattering process could also be a combination of the two afore-mentioned mechanisms for which the temperature dependence would be unknown. In principle, more information could be obtained from an evaluation of Eq. (5), which has not been done yet. The fits for x = 0.4 are the electron-impurity scattering as shown in **Figure 8**. **Figure 8a–c** show

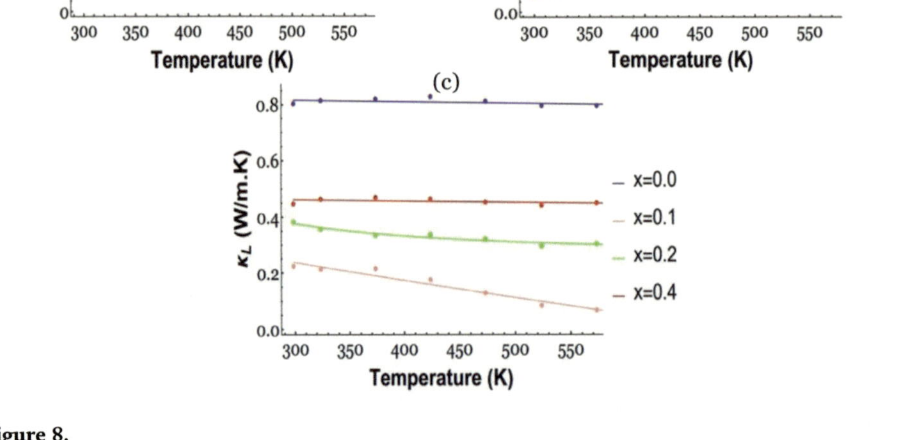

Figure 8.
Measured (a) electrical conductivity, (b) electronic-, and (c) lattice thermal conductivity fits as a function of T *for* $Pb_{(2-x)}Y_xRu_2O_{(6.5+z)}$ *with x = 0, 0.1, 0.2, and 0.4 moles of Y.*

$Pb_{(2-x)}Y_xRu_2O_{(6.5+z)}$ with x = 0, 0.1, 0.2, and 0.4 fitted with $T^{-1/2}$, $T^{1/2}$, and T^{-1} for σ, κ_e, and κ_L, respectively.

For $Pb_{(2-x)}Y_xRu_2O_{(6.5+z)}$ with x = 1.0 a clear selection of a mechanism cannot be made. The selection criteria, i.e., best least squares fit and goodness of the fit, allow for three scattering mechanisms. $\sigma(T)$ could vary with $T^{-1/2}$ and $\kappa_e(T)$ with $T^{1/2}$ or $\sigma(T)$ varies with T^{-1} and $\kappa_e(T)$ with T^0. Based on **Table 1** the underlying scattering mechanism would be electron-impurity scattering at $T \ll T_F$ or electron–phonon scattering at $T_{BG} \ll T \ll T_F$, or both (**Table 1**) [34]. The lattice thermal conductivity $\kappa_L(T)$ varies with T^{-1} implying Umklapp scattering and supports the electron-impurity scattering mechanism. The third possibility is that the Mott Variable Range Hopping (MVRH) mechanism [20] is active, which requires a $T^{1/2}$ dependence of $S(T)$ and for $\sigma(T)$ an $\exp(-1/T)^{1/4}$ dependence. Fitting $T^{1/2}$ to $S(T)$ yielded a satisfactory goodness of the fit. The same was the case when fitting the exponential function to the $\sigma(T)$ data. As I mentioned for x = 0.2 and 0.4, here with an yttrium content of 1.0 mole there could be a combination of the scattering mechanisms for which the temperature dependence could be determined using Eq. (5). This pyrochlore, like the ones in the previous section are in the MIT zone. This explains why there may be more than one scattering process active. The MVRH fitting is shown in **Figure 9**.

For $Pb_{(2-x)}Y_xRu_2O_{(6.5+z)}$ with x = 1.5 and 2.0 the best fits of electrical conductivity $\sigma(T)$ varies with $\exp(-1/T)^{1/4}$ and the Seebeck coefficients with $T^{1/2}$. As for x = 1.0, MVRH can be suggested. Mott's variable range hopping model describes hopping conduction between localized states with electron energies close to the Fermi level [41, 65, 66]. Hopping takes into account both thermally activated hopping over an energy threshold and phonon-assisted tunneling between localized states. For thermally activated hopping the Boltzmann factor $\exp(-W/k_BT)$ applies, where W is the energy barrier between the localized states. The electrical conductivity of the pyrochlores on the insulator side of the MIT shows thermally activated conduction characteristics (Figure 3 in [45]). This behavior indicates transport by the MVRH [67]. It could be due to a random potential caused by yttrium substitution [9]. **Figure 9** shows fitting of $Pb_{(2-x)}Y_xRu_2O_{(6.5+z)}$ with x = 1.0, 1.5, 2.0 by MVRH for σ and S.

The figure of merit zT of a thermoelectric material determines its value for practical applications. Respective values have been calculated for lead ruthenate derivatives and for Pb-Y ruthenate solid solutions. All zT values are still more than two orders of magnitude below those achievable today with other ceramics. The highest zT for all pyrochlores studied here was 7.3×10^{-3} at 523 K for $Pb_{1.9}Y_{0.1}RuO_{6.5\pm z}$. Since the focus of this work was on quantum physical

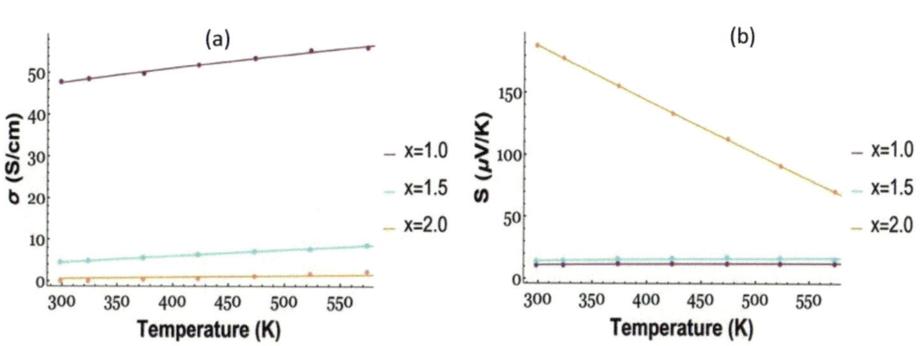

Figure 9.
Measured electrical conductivity (a) and Seebeck coefficients (b) with respective fits as a function of temperature for $Pb_{(2-x)}Y_xRu_2O_{(6.5+z)}$ with x = 1.0, 1.5, and 2.0.

interpretation of the pyrochlores' thermoelectric properties, no effort was made yet to search for pyrochlores with higher figures of merit.

7. Summary and conclusions

Thermoelectric properties of ceramics are determined by their crystal structure and chemical composition. Important structural details affecting the transport mechanisms of heat and electricity are, e.g., vacancies, impurities, lattice site occupancy, lone pair electrons, and substitutions. The nature of these structural details can determine or change scattering mechanisms that determine the thermoelectric performance of the material. A selection of published mechanisms has been compiled. The way of testing these mechanisms by fitting respective mathematical functions to experimental data and the quantitative evaluation of the fits have been shown in this chapter, based on a large number of experimental data.

All thermoelectric properties of isomorphic lead ruthenate pyrochlores (defect pyrochlores) with different Pb:Ru atom ratios were measured and analyzed in quantum-physical terms to interpret transport mechanisms of thermal and electrical conductivity and to understand the temperature dependence of the Seebeck effect. All seven pyrochlores were *p*-type, exhibiting common features such as site substitutions, vacancies and some impurities, to which transport mechanisms are sensitive. Therefore, the same scattering mechanisms were seen for all pyrochlores, specifically, electrical conductivity $\sigma(T)$, which varied with $T^{-1/2}$, the electronic part of thermal conductivity $\kappa_e(T)$ varied with $T^{1/2}$ and the lattice thermal conductivity $\kappa_L(T)$ with T^{-1}. Hence, $\sigma(T)$ and $\kappa_e(T)$ were governed by electron-impurity scattering and $\kappa_L(T)$ by the 3-phonon resistive process (Umklapp scattering), which supports the electron-impurity scattering mechanism. The Seebeck effect was inversely proportional to the carrier concentration.

Measurements and quantum-physical analyses were also done with lead-yttrium ruthenate solid solutions, which are defect pyrochlores as well. The temperature dependence of the Seebeck coefficient was qualitatively the same as for the lead ruthenate and derivatives. The same interpretation applies. A metal–insulator transition (MIT) occurred, if 0.2 moles of Pb were replaced by Y. The endmember $Pb_2Ru_2O_{6.5}$ shows metal-like behavior. $Y_2Ru_2O_7$ is an insulator. Impurity scattering prevailed until the MIT was reached. The temperature functions of electrical conductivity and electronic thermal conductivity were the same as for the lead ruthenates. On the semiconductor/insulator side of the MIT, up to about one mole of yttrium, several scattering processes were equally likely, e.g., electron-impurity- and electron–phonon scattering. From one to two moles of yttrium, the Mott-Variable-Range Hopping mechanism was active, as suggested by $\sigma(T)$ varying with $\exp(-1/T)^{1/4}$. All figures of merit zT were small, about 7.3×10^{-3} maximum at 523 K.

Based on these findings the following conclusions are provided:

- In previous work, glass-like thermal conductivity of $Pb_2Ru_2O_{6.5}$ was found. The same was the case here for lead ruthenate derivatives (Pb:Ru variations). Obviously, an increasing concentration of minority electrons with temperature, a likely explanation of glass-like behavior, is not significantly affected by the rather large changes of the pyrochlore stoichiometry and site occupation. Thermal conductivity increased at maximum only by a factor of less than two.

- Significant changes of electrical conductivity were seen and can be related to site occupancy and ion valence.

- Seebeck coefficients of $Pb_2Ru_2O_{6.5}$ increased slightly with temperature and decreased after a maximum at 423 K. A remarkable insensitivity of this temperature dependence vis-à-vis Pb:Ru changes was noticed, suggesting that the underlying temperature dependence of the concentration of minority electrons is little affected by Pb:Ru concentration changes.

- Mathematical analyses of the measured thermoelectric properties pointed at the same underlying scattering mechanisms in all compounds, mainly a result of structural defects, not of the Pb:Ru ratio.

- Introducing yttrium into the metal-like $Pb_2Ru_2O_{6.5}$ changes the thermoelectric properties more drastically than variations of the Pb:Ru ratio.

- Scattering mechanisms were affected by the metal–insulator transition (MIT). Changes of bond angles, bond distances, and electronic structures (Mott-Hubbard) at the MIT are likely to affect heat- and electricity transport, which was reflected by the finding that different scattering mechanisms were active on either side of the MIT.

- Knowledge of scattering mechanisms and their relation to structural details and composition can provide guidance to tailor and optimize thermoelectric properties of a material.

Acknowledgements

I would like to thank Dr. Werner Lutze, my former PhD adviser, for his constant support and consult and many discussions while I was writing this chapter. Many thanks go to Drs. Nicholas A. Mecholsky and David McKeown for discussions and Dr. Marek Brandys for building the instrument to measure electrical conductivity and Seebeck coefficients. I thank Dr. Ian Pegg for his encouragement and financial support, which made it possible for me to do this work.

Conflict of interest

The author does not have any conflict of interest.

Author details

Sepideh Akhbarifar
The Catholic University of America, Vitreous State Labratory, Washington D.C., USA

*Address all correspondence to: sepideha@vsl.cua.edu

References

[1] Terasaki I, Sasago Y, Uchinokura K. Large thermoelectric power in $NaCo_2O_4$ single crystals. Physical Review B. 1997; 56:R12685–R12687

[2] Chakoumakos B.C. Systematics of the pyrochlore structure type, ideal $A_2B_2X_6Y$. Journal of Solid State Chemistry. 1984; 53:120-129

[3] Chakoumakos B.C. Pyrochlore, McGraw-Hill Yearbook of Science & Technology 1987, edited by S.P. Parker, McGraw-Hill, Inc., New York; 1986. p. 393

[4] Subramanian M.A, Aravamudan G, Subba Rao G.V. Oxide pyrochlore–A review. Solid State Chemistry. 1983; 15: 55-143

[5] Sleight A.W, Bouchard R.J. National Bureau of Standards Special Publication 364, Solid State Chemistry, Proceedings, 5th Materials Research Symposium; 1972. p. 227

[6] Tachibana M, Kohama Y, Shimoyama T, Harada A, Taniyama T, Itoh M, Kawaji H, Atake T. Electronic properties of the metallic pyrochlore ruthenates $Pb_2Ru_2O_{6.5}$ and $Bi_2Ru_2O_7$. Physical Review B. 2006; 73:193107

[7] Mayer-von Kürthy G, Wischiert W, Kiemel R, Kemmler-Sack S. System $Bi_{2-x}Pb_xPt_{2-x}Ru_xO_{7-z}$: A pyrochlore series with a metal-insulator transition. Journal of Solid State Chemistry. 1989; 79:34-45

[8] Bouchard R.J, Gillson J.L. A new family of bismuth precious metal pyrochlores. Material Research Bulletin. 1971; 6:669-680

[9] Yoshii S, Sato M. Studies on Metal-Insulator transition of pyrochlore compound $Y_{2-x}Bi_xRu_2O_7$. Journal of the Physical Society of Japan. 1999; 68: 3034-3040

[10] Klein W, Kremer R.K, Jansen M. $Hg_2Ru_2O_7$, a new pyrochlore showing a metal-insulator transition. Journal of Materials Chemistry, 2007; 17: 1356-1360

[11] Takeda T, Nagata M, Kobayashi H, Kanno R, Kawamoto Y, Takano M, Kamiyama T, Izumi F, Sleight A.W. High-Pressure Synthesis, Crystal Structure, and Metal-Semiconductor Transitions in the $Tl_2Ru_2O_{7-\delta}$ Pyrochlore. Journal of Solid State Chemistry. 1998; 140:182-193

[12] Lee S, Park J.G, Adroja D.T, Khomskii D, Streltsov S, McEwen K.A, Sakai H, Yoshimura K, Anisimov V.I, Mori D, Kanno R, Ibberson R. Spin gap in $Tl_2Ru_2O_7$ and the possible formation of Haldane chains in three-dimensional crystals. Nature Materials. 2006; 5:471

[13] Lee K-S, Seo D-K., Whangbo M-H. Structural and Electronic Factors Governing the Metallic and Nonmetallic Properties of the Pyrochlores $A_2Ru_2O_{7-y}$. Journal of Solid State Chemistry. 1997; 131:405-408

[14] Hsu W.Y, Kasowski R.V, Miller T, Chiang T-C. Band structure of metallic pyrochlore ruthenate $Bi_2Ru_2O_7$ and $Pb_2Ru_2O_{6.5}$. Applied Physics Letter. 1988; 52:792

[15] Li L, Kennedy B.J. Structural and Electronic Properties of the Ru Pyrochlores $Bi_{2-y}Yb_yRu_2O_{7-\delta}$. Chemistry of Materials. 2003; 15:4060–4067

[16] Kennedy BJ, Vogt T. Structure and bonding trends in ruthenium pyrochlore. Journal of Solid State Chemistry. 1996; 126:261–270

[17] Takeda T, Kanno R, Kawamoto Y, Takeda Y, Yamamoto O. New Cathode Materials for Solid Oxide Fuel Cells Ruthenium Pyrochlores and Perovskites. Journal of the

Electrochemical Society 2000; 147: 1730-1733

[18] Beyerlein R.A, Horowitz H.S, Longo J.M, Leonowicz M.E, Jorgensen J. D, Rotella F.J. Neutron diffraction investigation of ordered oxygen vacancies in the defect pyrochlores, $Pb_2Ru_2O_{6.5}$ and $PbT_1Nb_2O_{6.5}$. Journal of Solid State Chemistry. 1984; 51:253-265

[19] Ishii F, Oguchi T. Electronic Band Structure of the Pyrochlore Ruthenium Oxides $A_2Ru_2O_7$ (A= Bi, Tl and Y). Journal of the Physical Society of Japan 2000; 69:526-531

[20] Cutler M, Leavy J.F, Fitzpatrick R.L. Electronic transport in semimetallic cerium sulfide. Phys. Rev. 1964; 133: A1143–A1152

[21] Grundmann M. The Physics of Semiconductors; An introduction including Devices and Nanophysics. Heidelberg: Springer, 2006

[22] Bhandari CM, Rowe DM. CRC Handbook of Thermoelectrics (ed. Rowe, D. M.) Ch. 5, 43–53 (CRC, Boca Raton, 1995)

[23] Neophytou N. Theory and simulation methods for electronic and phononic transport in thermoelectric materials, Springer, 2020

[24] Lukas K.C, Liu W.S, Joshi G, Zebarjadi M, Dressehaus M.S, Ren Z.F, Chen G, Opeil C.P. Experimental Determination of the Lorenz Number in $Cu_{0.01}Bi_2Te_{2.7}Se_{0.3}$ and $Bi_{0.88}Sb_{0.12}$. Physical Review B: Condensed Matter Materials Physics 2012; 85:205410

[25] Snyder G.J, Toberer E.S. Complex thermoelectric materials. Nature Materials. 2008; 7:105–114

[26] Thesberg M, Kosina H, Neophytou N. On the Lorenz number of multiband materials. Physical Review B. 2017; 95:125206

[27] Zhao L.D, Lo S.H, Zhang Y, Sun H, Tan G, Uher C, Wolverton C, Dravid V. P, Kanatzidis M.G. Ultralow thermal conductivity and high thermoelectric figure of merit in SnSe crystals. Nature. 2014; 508:373

[28] Völklein F, Reith H, Cornelius T.W, Rauber M, Neumann R. The experimental investigation of thermal conductivity and the Wiedemann-Franz law for single metallic nanowires. Nanotechnology. 2009; 20:325706

[29] Graf M.J, Yip S.K, Sauls J.A, Rainer D. Electronic thermal conductivity and the Wiedemann-Franz law for unconventional superconductors. Physical Review B. 1996; 53:15147

[30] Sun X.F, Lin B, Zhao X, Li L, Komiya S, Tsukada I, Ando Y. Deviation from the Wiedemann-Franz law induced by nonmagnetic impurities in over doped $La_{2-x}Sr_xCuO_4$. Physical Review B. 2009; 80:104510

[31] Thonhauser T, Scheidemantel T.J, Sofo J.O, Badding J.V, Mahan G.D. Thermoelectric properties of Sb_2Te_3 under pressure and uniaxial stress. Physical Review B. 2003; 68:085201

[32] Madsen G.K.H, Schwarz K, Blaha P, Singh D.J. Electronic structure and transport in type-I and type-VIII clathrates containing strontium, barium, and europium. Physical Review B. 2003; 68:125212

[33] Chaput L, Hug G, Pecheur P, Scherrer H. Anisotropy and thermopower in Ti_3SiC_2. Physical Review B. 2005; 71:121104(R)

[34] Lavasani A, Bulmash D, DasSarma S. Wiedemann-Franz law and Fermi liquids. Physical Review B. 2019; 99:085104

[35] Beekman M, Cahill D.G. Inorganic Crystals with Glass-Like and Ultralow Thermal Conductivities. Crystal

Research and Technology. 2017; 52: 1700114

[36] Sathish S, Awasthi O.N. Electron-Electron scattering and low-temperature electrical resistivity in copper and silver. Physics Letter A. 1984; 100:215-217

[37] Yildiz A, Serin N, Serin T, Kasap M. Crossover from Nearest-Neighbor Hopping Conduction to Efros–Shklovskii Variable-Range Hopping Conduction in Hydrogenated Amorphous Silicon Films. Japanese Journal of Applied Physics. 2009; 48:111203

[38] Ashcroft N.W, Mermin N.D. Solid State Physics. Saunders College Publishing, Fort Worth. 1976. p. 657

[39] Sharma S.K, Sagar P, Gupta H, Kumar R, Mehra R.M. Meyer-Neldel rule in Se and S-doped hydrogenated amorphous silicon. Solid State Electronics. 2007; 51:1124-1128

[40] Zvyagin I.P, Kurova I.A, Ormont N. N. Variable range hopping in hydrogenated amorphous silicon. Physica Status Solidi C. 2004; 1:101

[41] Mott N.F, Davis E.A. Electronic Properties in Non-Crystalline Materials (Clarendon Press, Oxford, U.K., 1971)

[42] Efros A.L, Shklovskii B.I. Electronic Properties of Doped Semiconductors (Springer, Berlin, 1984)

[43] Zabrodskii A.G, Zinoveva K.N. Low-temperature conductivity and metal-insulator transition in compensate n-Ge. Journal of Experimental and Theoretical Physics. 1984; 59:425

[44] Liu H, Choe H.S, Chen Y, Suh J, Ko C.h, Tongay S, Wu J. Variable range hopping electric and thermoelectric transport in anisotropic black phosphorus. Applied Physics Letter. 2017; 111:102101

[45] Akhbarifar S, Lutze W, Mecholsky N.A, Pegg I.L. Metal-insulator transition in lead yttrium ruthenate. Materials Chemistry and Physics. 2021; 260:124172

[46] Parrondo J, George M, Capuano Ch, Ayers K.E, Ramani V. Pyrochlore elec trocatalysts for efficient alkaline water e lectrolysis. Journal of Materials Chemistry A. 2015; 3:10819-10828

[47] Song K-W., Lee K-T. Characterization of $Pb_2Ru_{2-x}Bi_xO_7$ (x= 0, 0.2, and 0.4) Pyrochlore Oxide Cathode Materials for Intermediate Temperature Solid Oxide Fuel Cells. Journal of Ceramic Processing Research. 2011; 12:30-33

[48] Akhbarifar S, Lutze W, Mecholsky N.A, Pegg I.L. Glass-like thermal conductivity and other thermoelectric properties of lead ruthenate pyrochlore. Materials Letters. 2020; 272:128153

[49] Sleight A.W. High pressure synthe sis of platinum metal pyrochlores of the type $Pb_2M_2O_{6-y}$. Materials Research Bulletin 1971; 6:775-780

[50] Kobayashi H, Kanno R, Kawamoto Y, Kamiyama T, Izumi F, Sleight A.W. Synthesis, Crystal Structure, and Electrical Properties of the Pyrochlores $Pb_{2-x}Ln_xRu_2O_{7-y}$ (Ln= Nd, Gd). Journal of Solid State Chemistry. 1995; 114:15-23

[51] Paschen S, Chapter 15 (Thermoelectric Aspects of Strongly Correlated Electron Systems), in Thermoelectrics Handbook (Editor: Rowe D.M.). 2006

[52] Horowitz H.S, Longo J.M, Lewandowski J.T. New oxide pyrochlores: $A_2[B_{2-x}A_x]O_{7-y}$ (A = Pb, Bi; B = Ru, Ir). Materials Research Bulletins. 1981; 16:489–496

[53] Hwang H.Y, Cheong S-W. Low-field magnetoresistance in the pyrochlore $Tl_2Mn_2O_7$. Nature. 1997; 389:942-944

[54] Bramwell S.T, Gingras M.J.P. Spin ice state in frustrated magnetic pyrochlore. Science. 2001; 294:1495-1501

[55] Sidharth D, Alagar Nedunchezhian A., Rajkumar R, Yalini Devi N, Rajasekaran P, Arivanandhan M, Fujiwara K, Anbalaganb G, Jayavel R. Enhancing effects of Te substitution on the thermoelectric power factor of nanostructured $SnSe_{1-x}Te_x$. Physical Chemistry Chemical Physics. 2019; 21: 15725-15733

[56] Tse J.S, Uehara K, Rousseau R, Ker A, Ratcliffe CI, White MA, MacKay G. Structural Principles and Amorphous like Thermal Conductivity of Na-Doped Si Clathrates. Physical Review Letter. 2000; 85:114-117

[57] Damodara Das V, Soundararajan N. Size and temperature effects on the Seebeck coefficient of thin bismuth films. Physical Review B. 1987; 35: 5990-5996

[58] Goldsmid H.J, Sharp J.W. Estimation of the thermal band gap of a semiconductor from Seebeck measurements. Journal of Electron Materials 1999; 28:869-872

[59] Beekman M, Cahill D.G. Inorganic Crystals with Glass-Like and Ultralow Thermal Conductivities. Crystal Research and Technology. 2017; 52: 1700114

[60] Skoug E.J, Morelli D.T. Role of Lone-Pair Electrons in Producing Minimum Thermal Conductivity in Nitrogen-Group Chalcogenide Compounds. Physical Review Letter. 2011; 107:235901

[61] Zhang Y, Skoug E, Cain J, Ozoliņš V, Morelli D, Wolverton C. First-principles description of anomalously low lattice thermal conductivity in thermoelectric Cu-Sb-Se ternary semiconductors. Physical Review B. 2012; 85:054306

[62] Nielsen M.D, Ozolins V, Heremans J.P. Lone-pair electrons minimize lattice thermal conductivity, Energy Environ. Science. 2013; 6: 570-578

[63] Dong Y, Khabibullin A.R, Wei K, Salvador J.R, Nolas G.S, Woods L.M. Bournonite $PbCuSbS_3$: stereochemically active lone-pair electrons that induce low thermal conductivity. ChemPhysChem. 2015; 16:3264 -70

[64] Schelling P.K, Phillpot S.R. Mechanism of Thermal Transport in Zirconia and Yttria-Stabilized Zirconia by Molecular-Dynamics Simulation. Journal of American Ceramic Society. 2001; 84:2997

[65] Shklovskii B.I, Efros A.L. Electronic Properties of Doped Semiconductors. Springer, London. 1984

[66] Park T.E, Suh J, Seo D, Park J, Lin D. Y, Huang Y.S, Choi H.J, Wu J, Jang C, Chang J. Hopping conduction in p-type MoS_2 near the critical regime of the metal-insulator transition. Applied Physics Letter. 2015; 107:223107

[67] Mott N.F. The metallic and Nonmetallic States of Matter. (P. P. Edwards and C. N. R. Rao, Eds.), p. 1, Taylor and Francis, London/ Philadelphia. 1985